THE VIRGIN AND THE DYNAMO

ROBERT ROYAL is vice president for research and a senior fellow in religion and society at the Ethics and Public Policy Center, Washington, D.C. He is the author of the Center's *1492 and All That: Political Manipulations of History* and of *Dante* in the Crossroads "Spiritual Legacy" series, and he has edited or co-edited nine other books, including *Building the Free Society* (with George Weigel) and *The Hospitable Canon* (with Virgil Nemoianu). In addition, he is the translator of Jean-Pierre Torrell's *Thomas Aquinas: The Person and His Work* and Roberto Papini's *The Christian Democrat International*. Among the numerous publications in which his articles have appeared are the *Washington Post, First Things, Crisis,* the *Catholic Historical Review, Communio,* and the *American Enterprise Magazine.*

THE VIRGIN AND THE DYNAMO

Use and Abuse of Religion
in Environmental Debates

ROBERT ROYAL

ETHICS AND PUBLIC POLICY CENTER
WASHINGTON, D.C.

WILLIAM B. EERDMANS PUBLISHING COMPANY
GRAND RAPIDS, MICHIGAN

Copyright © 1999 by the Ethics and Public Policy Center
1015 Fifteenth St. N.W., Washington, D.C. 20005

Published jointly 1999 by the Ethics and Public Policy Center and
Wm. B. Eerdmans Publishing Co.
255 Jefferson Ave. S.E., Grand Rapids, Mich. 49503

Printed in the United States of America

04 03 02 01 00 99   7 6 5 4 3 2 1

Library of Congress Cataloging-in-Publication Data

Royal, Robert, 1949-
The Virgin and the dynamo: use and abuse of religion
in environmental debates / Robert Royal.
p.   cm.
ISBN 0-8028-4468-5 (pbk.: alk. paper)
1. Ecology — Religious aspects — Christianity.
2. Ecology — Religious aspects.   I. Title.
BT695.5.R69   1999
291.1′78362 — dc21   98-50276
CIP

What born fools were all who lived in ignorance of God! From the good things before their eyes they could not learn to know him who is, and failed to recognize the artificer though they observed his handiwork! Fire, wind, swift air, the circle of the starry signs, rushing water, or the great lights in heaven that rule the world—these they accounted gods. If it was through delight in the beauty of these things that people supposed them gods, they ought to have understood how much better is the Lord and master of them all; for it was by the prime author of all beauty they were created. If it was through astonishment at their power and influence, people should have learnt from these how much more powerful is he who made them. For the greatness and beauty of created things give us a corresponding idea of their Creator.

Wisdom of Solomon 13:1-5

Contents

Acknowledgments

L ike many people of a certain age, I first had my eyes opened to care for nature and a natural sense of the sacredness of creation by watching the way parents, grandparents, and—fortunately for me —even great-grandparents acted and spoke. Those generations knew how to persuade the rocky soil and troublesome climate of New England to produce not only traditional fruits like apples, pears, and plums but also more exotic and delicate harvests of peaches, grapes, figs, herbs, vegetables, and flowers. For them, there was never any necessary conflict between respect for nature in itself and the natural utility and beauty they knew how to enhance. At the same time, they never fell into mere sentimentality about the world or underestimated the sheer backbreaking work needed to tame it, guide it, or merely live with it. Several of them came to the United States as immigrants who were grateful for the opportunities they found here to escape political repression and the hard lives on the land they had lived in Europe. This entire book is in many ways only a poor verbal shadow of a rich and varied reality that they lived.

Among my other teachers, my brother, the now Rev. Kevin Royal, comes first. When we were boys, we spent every second possible in woods that have since disappeared to development. But his early piety will always be associated in my mind with certain fields and ponds, trees and slopes. Years later, and long before concern for wetlands became fashionable, it was under his guidance in his canoe that my wife and I discovered that what we thought was merely a smelly shoreline near a regional airport in my home town was in fact a marvelous salt marsh inhabited by giant blue herons, snowy egrets, snapping turtles (dining on baby ducks that day), and many other wonders. It was and is typical of him to find such splendors where others do not have enough faith to look.

Many other people have contributed directly and indirectly to this work. My own family put up bravely with my absence and preoccupation over this book on and off over several years. My wife, Veronica, was a strong inspiration, spiritually and in every other way. Elizabeth, John Paul, and Natalie were a concrete reminder of who will have to live in the future with the decisions we are making today. I only pray the product is worth the sacrifice.

My old friend Ambassador José Osvaldo de Meira Penna, now retired from the Brazilian Foreign Service, has long enlightened me about so many issues that to thank him here merely for his help in thinking about the environment and politics in his country would border on ingratitude. We have not yet had the chance to visit the Amazon together, but I benefitted in Rio de Janeiro from his company on a drive through the *floresta de Tijuca,* the only rain forest within the limits of a major city.

Jason Boffetti was as good a research assistant, in this as in so many other projects, as anyone could want. Kristina Fox helped in the final preparation of the manuscript. Diane Bryhn helped clarify some language and cultural questions about the Norwegian Deep Ecologist Arne Naess. George Weigel and Elliott Abrams were patient beyond the call of duty as this book slowly took shape

Heartfelt thanks are especially due to the Earhart Foundation for crucial financial support early in the project.

Ron Bailey, Jody Bottum, Todd Breyfogle, Stratford Caldecott, Samuel Casey Carter, Thomas Sieger Derr, Russell Hittinger, Deal Hudson, Andrew Kimbrell, the Rev. Francis Martin, David Murray, Robert Nelson, David Schindler, Christina Hoff Sommers, Scott Walter, and my colleagues at the Ethics and Pubic Policy Center, Michael Uhlmann and Michael Cromartie, have at various stages in the process and in different ways done me the favor of suggesting other lines of inquiry, correcting my errors and ignorance, encouraging me, or just usefully disagreeing. My editor Cheryl Hoffman not only kept my language properly disciplined but also added a great deal at many points to my arguments. Carol Griffith also provided valuable editorial advice.

I owe particular thanks to Christopher DeMuth, president of the American Enterprise Institute, for asking me to give one of AEI's Bradley lectures in 1997, which presented an opportunity to further sharpen my thinking.

For other help provided, my heartfelt thanks to Ginny Berzin, Diane Fleming, Janet Hillier, Cornelia Sgoutas, the Poor Clares of Alexandria, and the indispensable "P.P."

Trying to figure out how God, nature, and human beings relate to one another so as to preserve the spiritual goals of the Creator and to meet the many legitimate needs of creatures has been one of the most difficult tasks I have ever undertaken. Anyone with even a modicum of self-knowledge knows that the question exceeds his personal powers. Even the human species as a whole has to content itself with limited approaches to what is ultimately a great mystery. If this book enables people, with or without religious beliefs, to appreciate more fully the complexities of every dimension of this problem, it will have done as much as any writer on this subject can reasonably hope.

September 8, 1998
Washington, D.C.

INTRODUCTION

The Virgin and the Dynamo

In and around New York City, there are Buddhist communities that believe compassion towards all living beings requires them to set captive animals free into nature. Some people buy goldfish and release them in lakes, others allow caged finches and parakeets to fly off, still others buy turtles in shops and seek out suitable habitats where they can live in the wild. All are obeying an ancient Sutra in Buddhist scriptures going back to the fifth century: "If you see a person about to kill an animal, you should devise a means to rescue and protect that creature." Strict vegetarians, they regard all living things as equally valuable. Though many modern Westerners might think other issues more pressing, it would be difficult to imagine a more sincere religious response to the suffering or potential death of even the humblest living creatures.

Yet these compassionate acts have results that are often quite different from those intended. For example, releasing goldfish into lakes in the New York City area disturbs the habitats of indigenous species. The chief of the New Jersey Bureau of Freshwater Fisheries has said that the goldfish are so prolific that "they have an edge over the natural population. . . . They overpopulate and kill off the perch, sunfish, catfish, and everything else."[1] Similarly, naturalists have remarked that releasing captive birds into the wild is in most cases a form of killing with kindness, since domesticated birds are unable to cope with natural predators and other environmental threats. And setting free commercially available turtles may have the worst environmental effect of all; many of

1

them carry diseases that can affect already threatened indigenous species of turtles.[2]

Buddhism gets particularly high marks in environmental literature because its principles of universal harmony and compassion towards all sentient beings are regarded as sound starting points for the whole environmental movement. If those principles are properly understood, that may be true, though it is a lot easier to get a grasp on compassion than on harmony, as the anecdote itself indicates. Strictly speaking, these Buddhists were not thinking about "the environment"; they were only trying to save animals. And a good case could be made that serious attention to local problems by individuals and small groups could be a highly effective way of caring for the world. But such assumptions do not get us very far, and not only for Buddhism. Christians and Jews, too, find in their Scriptures God's command to be stewards over the earth, which may be understood to entail wide-ranging responsibilities. That bare fact, however, does not give us much concrete guidance in what, if anything, to do about greenhouse gas emissions, wetland preservation, endangered species, the ozone layer, or a host of other issues that often involve conflicting values or inescapable trade-offs.

The usual religious approach to environmentalism today, even among quite sophisticated people, takes as its point of departure the Book of Genesis's proclamation of the goodness of creation and compares it with the horrors, real and alleged, of what we humans have done to the earthly Paradise. As a guiding framework, there is something to be said for this image, if we use it carefully. Unfortunately, that is rarely the case. One typical book by an otherwise quite sophisticated writer, for example, asserts that the environment kills 35,000 children daily in the developing world. Pollution controls are lax or ineffective in developing nations, the author says, and human greed and perversity are responsible for those 35,000 deaths.[3] Perhaps so, but not exactly the perversity that the author suggests. As anyone who knows the Third World could explain, most childhood deaths there (about a million a year) are due to the *lack* of industrialization and development, which leaves children vulnerable to diarrhea and dehydration from both polluted and "naturally" unsafe water; to respiratory problems from the smoke from organic fuels, mostly dung and wood, burnt in huts; and to the numerous other rigors of unimproved nature itself.

Why God decided to make an earth that contains natural dangers for us and requires no little effort and knowledge, patiently accumulated over the ages, to arrive at human flourishing, remains a mystery. But if there is anything that modern science has demonstrated about the earth it is that the state of nature, abstractly considered, is neither a natural paradise nor even, in any simple or self-evident way, in harmony and stability. To begin with, the forms of nature we see around us are the product of several billion years of development on earth. As a species, we need to become far more conscious of the ways in which we may damage or even destroy whole sections of this priceless patrimony. The other side of this same realization, however, is that nature has always been changing, sometimes in gentle increments, sometimes with brutal force.

Natural Challenges

The most recent Ice Age, for example, ended about twelve thousand years ago. Prior to that, glaciers miles thick covered the northern parts of Europe and the Americas. In what is today the United States, the northern forests of which we are so proud had been literally scraped off the face of the earth. Naturally, few if any animals survived in these regions. All of Canada and New England, Minnesota, the Dakotas, Montana, Wyoming, Idaho, Oregon, Washington — some of the most stunning parts of the North American landscape — regrew from scratch as the ice receded. It is worth remembering these natural disasters and recovery whenever we encounter some statement about the "irreversibility" of some environmental trend. Nature has destroyed and rebuilt large portions of itself in the past on a scale that human beings have not matched and, in all probability, never will.

Or take variations in global temperatures, which we rightly worry may have something to do with anthropogenic (i.e., human-caused) changes in the concentration of greenhouse gases in the atmosphere. Long before human activity could have had any measurable effect on them, global temperatures fluctuated sharply and sometimes rapidly. For instance, around nine thousand years ago, it appears, we experienced the peak temperature of what scientists call our interglacial period, about four degrees Fahrenheit higher than the present. In other words, in the recent past, as geological time goes, nature itself produced temperatures near the high estimate in the worst-case scenario for global warming without human intervention. This fact does not mean that our industrialized civilization can rest easy in its consumption of fossil fuels, but it places the usual doomsday warnings in a different perspective based on careful attention to the world God created rather than the one a fundamentalist reading of the Bible or a scientifically uninformed view might imagine he created.

Nor are such occurrences limited to prehistoric times. Around the year 1000, a period of warming occurred that appears to have made crossing the Atlantic easier and may have had something to do with the fact that Leif Eriksson found grapes growing in the northern land he called Vineland. By contrast, from perhaps 1550 to around 1850, the earth went through a "Little Ice Age" when temperatures were lower than average. As we shall see, volcanic eruptions periodically caused years "without a summer," during the same period. Part of the difficulty in determining whether global warming is caused by human production of carbon dioxide (CO_2) is that, when the cool period ended, global temperatures started to rise. Seventy per cent of the temperature increase in the last century or so occurred prior to 1940, which is to say before the industrialization that led to the big increases of CO_2.

But perhaps we should take an even longer time perspective. The Bible and most of the great religious traditions of the world affirm that God, or some source of goodness, willed the whole universe into existence and continues to watch over it. But when we begin to look into the specific nature of that providence, some troubling facts arise. First, while creation as we understand it today involves the evolution of vast and complex physical and biological systems, it also involves the destruction of much along the way. All the elements except for the very lightest (hydrogen, helium, and lithium) were cooked up from the lighter elements within earlier generations of stars. These stars exploded and sent those elements, including those that make up our own bodies, out into the universe, where they have assembled into various configurations. So the natural order we see around us is the result of cosmic creation and destruction.

Nor is this creation and destruction limited to non-living things. The best scientific evidence indicates that roughly 99.5 per cent of all the species that ever existed on earth are now extinct, all but a tiny fraction from "natural" events that occurred long before human beings had even come into the cosmic picture. The destruction of species and the generation of new ones have been going on in cycles virtually since the appearance of life on the planet, sometimes at speeds far exceeding even the highest estimates of human-caused extinctions in our century. The very oxygen in our atmosphere, on which we rely for life, was originally the waste product of a micro-organism that became so prolific that the highly reactive gas it produced almost killed off all other living things. To put this in biblical terms, a modern Noah might be able to save all existing species over the short run, but the forces that gave rise to those species will, in the short run by geological standards, make saving all species impossible without massive human intervention in "natural processes."

But what of the vast forces that human beings are now able to wield against nature, and the pollutants we add to our pristine environment? The fact that we put the question this way is already culturally significant. The three hundred years or so in which the scientific culture of the West expanded without much concern about its effects on nature may now be viewed as a period that was badly one-sided. But the one-sidedness was more a matter of innocence than of deliberate harm: all cultures have affected nature within their capacity to do so. The specifically modern difficulty has been that science since the seventeenth century, often against the very intentions of men like Kepler, Copernicus, and Galileo, seemed to reduce nature to mere energy and resources. Since the explosion of the old religious universe by science seemed irreversible, we were told that we would all have to submit and get used to a cold universe in which we ourselves created meaning. The various modern ideologies — Marxism, Freudianism, existentialism, humanism — all seemed to be based on the truth of that proposition. But a curious thing happened as we advanced further in this world.

First, we began to notice that we ourselves, we science-making, symbol-mongering, metaphor-producing, meaning-hungry beings, were also part of the cosmic evolution. It might almost appear to an impartial observer that the curious rise towards complexity on earth was one of the goals that nature had been striving towards for billions of years. Furthermore, the various ways of consciously appreciating the universe as something more than matter in motion (though also inescapably that) mean that the universe, in a sense, was recapitulating itself as awareness in the skulls of a higher primate on a planet of otherwise ordinary importance. In classical philosophy and traditional religious understanding, this human difference was thought of as something almost godlike. Both of those universes of thought found matter and mind inextricably connected, though never merely reducible, to each other. It was only when the overall culture began to regard matter alone as real and mind as a kind of inexplicable leftover that the whole problematic of human against nature, of the pristine opposed to the incessantly active and indeterminate, took on the modern form it has.

Look at the cultural frame we have imposed on power over nature. The poet Frederick Turner has observed that if the Grand Canyon had been produced by a vast strip-mining operation instead of by natural processes, many of us would regard it as a "scar upon the land."[4] Yet it could have been created through a vast human project lasting probably several hundred years or more — far less time than natural forces would have taken. And if we look at either path to the Grand Canyon, there seems nothing of great concern to nature in it, even allowing for the difference in time scales. The human version, like the natural one, would have had many and complex effects on the environment for hundreds of miles around, but it would have not much altered the basic patterns of earth. We find the canyon beautiful, as we would not find the mine, because of our assumptions about natural beauty.

Certain types of environmentalists, Al Gore prominently among them, like to point to human foolishness in projects such as the Soviet diversion of river water for the irrigation of cotton fields, which led to the virtual drying up of the Aral Sea, as evidence of our unprecedented impact.[5] Yet shortsighted as it was, the Soviet action only did what natural forces at many times and places throughout geologic time have done. Geologists believe, for example, that what we now call the Mediterranean Sea became temporarily cut off from other oceans about six million years ago. In less than one thousand years, the whole sea evaporated, leaving only a salt basin, until the oceans broke in again and refilled it rather rapidly. The process may have been repeated several times on the Mediterranean.[6] Again, the rate of change is more dramatic in the case of the human-caused emptying of the Aral, and we certainly do not want to repeat such mistakes very often, for good ecological as well as aesthetic reasons. But nature itself has a vast repertoire of change, and it would almost take a deliberate and sustained effort on our part, with a few exceptions, to exceed its own variabilities.

Learning from Nature

Two main lessons can be extracted from this scientific gloss on a religious view of the cosmos: First, the human species can, as some of our biological predecessors have, destroy itself and other species while unconsciously pursuing what seem to be goods. But, second, the kind of balance of nature often spoken of in environmental literature and the proclamation of the goodness of creation in religious environmentalism are more properly described as one set of earthly conditions set within much larger dynamics. Many aspects of the nature we wish to preserve are adapted to our health and well-being. And as great religious figures throughout history have demonstrated over and over again, contemplation of the natural order in its recurrences — and radical transience — can lead us to living contact with divine wisdom. Those of us who take our spiritual bearings from the Bible have deep responsibilities towards creation, but we must also keep in mind the Bible's frequent warnings about two perennial temptations: making an idol of nature and neglecting the needs of our human neighbor.

Rather than viewing human and environmental costs as intrinsic to the industrialization of the West and later the rest of the world, we might do better to regard them as a necessary, but partly transitory, phase in human history. The sweatshops and workhouses depicted in Dickens, terrible as they were, led on to the point where industrial workers in advanced societies live lives unattainable by kings in the past. Similarly, the industrial processes that once belched black smoke into the air, poured wastes into the waters, and buried dangerous materials in unsuitable places are rapidly turning environmentally friendlier and will become even more so, by human decision, in the near future.

All this is very far from the typical approach of environmentalists, religious among them, who—science notwithstanding—seem to believe that the only changes that occur in the biosphere are caused by wicked multinational corporations or First World consumers or Third World masses who insist on having children. A type of fundamentalism about the goodness of creation has obscured an older and far more realistic view. To put it in scriptural terms, most religious thinkers about environmental issues are far too mesmerized by the idea of a return to the Garden, a return that the Bible gives us little reason to expect, and too little attracted by the idea of a New Jerusalem, for which the Scriptures encourage us to strive. In this perspective, the human dominion over creation spoken of in Genesis is all hubris and foolishness compared with the state of nature. The consequences of this view are not long in appearing. It is not at all rare to find in religious environmental literature, for example, the claim that "nature would be better off without the human race," an idea that God and nature alike seem to have decided against long before we ever appeared. But in light of the nature that we see all around us, we have to recognize that, while human action may lead to the destruction of the race and the environment, so may human inaction.

Current quarrels over the environment, then, like modern quarrels over many other things, may be better seen as really about something quite different. As a consequence, they are unlikely to be resolved by more rigorous reasoning or by greater emotional sensitivity or even by facts. The debate, such as it is, shows a reasonable amount of rationality once the premises of the debaters are granted. The passionate environmentalist does not criticize the developmentalist's logic but accuses him of profound blindness to the damage that developed industrial societies inflict on animals, plants, and inanimate nature. The ardent developmentalist returns the compliment by pointing to the environmentalist's elaborate and quite rational concern for whales, baby seals, and rain forests and inhumane indifference to the still massive needs of his own species. The resulting stalemate suggests that deep issues of a different nature are lurking behind our disagreements.

No Theology, Please, We're American

All questions are ultimately theological — at least for theologians. The rest of us tend to think we can conduct politics, the intellectual life, science, business, art, and any number of other worldly pursuits without having to drag God into them. In the first place, it goes against our pragmatic bent in America to suppose otherwise. Though we routinely ask God to bless America on ceremonial occasions, the rest of the time we seem to believe that human effort and human understanding conceived along scientific lines are what make the world work. But even more important, we pluralistic Americans have the vague sense that if we have to settle theological disputes before we can tackle environmental issues, we will never get around to doing anything for man or nature.

I am not a theologian, and I do not think it would be wise to spend much time reading modern theology for enlightenment on environmental or other human problems. For the most part, modern theology (with a few notable exceptions) seems to me far more likely to make us even more dazed than we already are as we approach the end of the second Christian millennium. Theology is not the only discipline in this predicament. Philosophy, politics, literature, history, and all the traditional liberal arts — the disciplines that were supposed to liberate us from our slavery to ignorance and impulse — seem long on methods and theories, short on insight and wisdom. Most modern theologies are part of a larger cultural crisis that, for lack of a better term, is sometimes called postmodern. As the very name shows, postmodernism is a condition in which we seem suspended between the old "modern" view of the world and something that we cannot as yet name more clearly.

Yet modern theology is not the whole of theology. And it may be precisely what many people have discarded as old, outmoded, and useless that will provide us with a broader vision and a more fruitful point of departure for evaluating environmental and other human challenges. More than a century ago, in the heyday of progress, T. H. Huxley, the notorious defender of Darwin and inventor of the term *agnosticism,* confidently asserted in a review of the *Origin of Species:* "Extinguished theologians lie about the cradle of every science as the strangled snakes beside that of Hercules; and history records that whenever science and orthodoxy have been fairly opposed, the latter has been forced to retire from the list, bleeding and crushed if not annihilated, scotched, if not slain."[7] For several generations, that progressive view of science and its offspring, technology, as heroes overcoming the serpentine coils of ignorance and superstition while bringing great benefits to the human race seemed to some people unshakable. But the very magnitude of our success has made scientific worldviews and technology-dominated societies look somewhat serpentlike themselves and has led to a crisis of confidence in everything that we once thought held unlimited promise.

These twin crises — in both humane learning and the sciences — have resulted in a novel situation. We live longer and with greater abundance than ever before but fear that disasters of apocalyptic proportions lie just around the corner. Aldous Huxley, a descendant of T. H.'s, wrote *Brave New World* (1932) to warn about some alarming trends in modern societies. Literature is not always a good guide to scientific or social truths, but the latter Huxley touched an authentic nerve. He gave powerful expression to the truth that scientific advance had unintended and, often enough, troubling consequences. No contemporary promoter of science, not even Carl Sagan, has been as confident about the sheer rightness of scientific progress and the mere wrongness of prescientific views as were his nineteenth-century counterparts. Indeed, one of the striking things about our time is that we often see scientists and scholars in other modern disciplines referring us back, usually somewhat vaguely, not only to some traditional religious notions but also to the presumed wisdom of primitive peoples.

The Prescience of Henry Adams

In a curious way, the great American historian Henry Adams, author of the celebrated *Mont Saint-Michel and Chartres,* anticipated our current dilemma. In chapter 25 of his autobiography, *The Education of Henry Adams,* he characterized our age as torn between the Virgin and the Dynamo. The Virgin, an image of the fullness of religious belief and human meaning as well as beauty and nature itself, represented a spirituality that Adams deeply regretted he could no longer accept. Adams thought the Dynamo, the efficient and powerful achievements of modern science and technology, had forever destroyed the plausibility of truths associated with the Virgin. The Dynamo had, in fact, become an alternative religion, unrecognized as such by most people, but dominant in its power and effects all the same.

Adams was deeply ambivalent about this development. On the one hand, he appreciated the force and even the poetry of the Dynamo and lamented that his mostly literary education had not prepared him to deal with the world the Dynamo was bringing into existence. On the other hand, he felt that the absence of the Virgin meant the loss of something great and irreplaceable. The Virgin belonged to the ages of "superstition," but Adams could not wholly give her up. Indeed, this pessimistic son of one of America's most distinguished Protestant families appears to have said secret prayers to her in his despair about the future. "Prayer to the Virgin of Chartres," a poem found sewn inside his clothing at his death, casts American history and the Enlightenment project as self-destructive:

> Crossing the hostile sea, our greedy band
>   Saw rising hills and forests in the blue;
> Our father's kingdom in the promised land!
>   —We seized it, and dethroned the father too.

> And now we are the Father, with our brood,
>   Ruling the Infinite, not Three but One;
> We made our world and saw that it was good;
>   Ourselves we worship, and we have no Son.

Adams looks to the Virgin for consolation:

> So, while we slowly rack and torture death
>   And wait for what the final void will show,
> Waiting I feel the energy of faith
>   Not in the future science, but in you![8]

Despite all Adams's skepticism, the Virgin's combination of beauty, sex, goodness, nurturing, and organic power still strongly drew his modern soul: "at the Louvre and at Chartres . . . was the highest energy ever known to man, the creator of four-fifths of the noblest art, exercising vastly more attraction over the human mind than all the steam-engines and dynamos ever dreamed of; and yet this energy was unknown to the American mind. . . . All the steam in the world could not, like the Virgin, build Chartres."[9]

As penetrating as this analysis was and is, Adams's agonized sense that the Dynamo had irrevocably displaced the Virgin may have been premature. Besides his presumption that religion was simply finished (an assumption that appears less and less likely as we approach the end of the twentieth century), Adams's analysis overlooks the ways in which nature — the vast expanse of forests, cataracts, and mountains discovered by the settlers of the New World — served as a substitute Virgin in the American psyche for many years, and continues to do so today. America and Europe shared that perception of the American wilderness at the start of the modern era. European Romantics like Rousseau and Chateaubriand and a variety of figures like Emerson and Thoreau in the United States found in pristine nature a kind of functional equivalent of the Virgin. The American wilderness was sublime, a pure order that rebuked the human order.

Enter John Muir

In John Muir, one of the founders of the Sierra Club and a tireless evangelist for the virtues of sheer wildness, Romantic notions of nature arrived at a new high-water mark. Muir would not have cared much for the historical Virgin, who, in his strict Scots Calvinist upbringing, was regarded as a Roman Catholic superstition. But he cared passionately for what he called "virgin forests" that contained "thousands of God's wild blessings."[10] He had learned the religion of nature from Emerson and Thoreau, as well as from the American wilderness ("the University of the Wilderness"). It may never have built a Chartres, but it did not need to. For Muir, forests themselves were "God's First Temples."[11]

At times, Muir's veneration of untouched nature seems simply to cast all things human, particularly civilization, into the role of sheer error and even sinfulness. Muir had escaped the harsh antinature Christianity of his Scots Calvinist father through remarkable experiences that led him to venerate the wild. But he brought to the new religion the same prophetic fervor, antihuman bent, and moral absolutism that his ancestors had brought to the ancient faith.[12] For him, what was human was corrupt, as only a Calvinist can construe corruption, and the human was corrupting nature. In line with the Calvinist belief that the righteous state was a curb on sin, Muir had high hopes for direct political remedies for environmental problems. Public ownership of land had to be superior to private ownership. All human activity — forestry, animal grazing, and other useful commercial enterprises — only desecrated his holy of holies. All the cultural achievements that had been patiently accumulated through the ages were so many obstacles to authentic human life: "Civilized man chokes his soul, as the heathen Chinese their feet."[13]

Anticivilization views, of course, have a long pedigree in Western thought. Rousseau, the originator of much of the noble-savage myth, while writing his tract on education, *Emile,* had made a similar point almost 150 years before Muir: "It is absolutely certain that the learned societies of Europe are but so many public schools of falsehood; and very surely there are more errors in the Academy of Sciences than in the whole tribe of Hurons."[14] In this new dispensation, human society with its slow accretion of "unnatural" advances in science, commerce, and learning seems like a variegated and false encrustation on a more basic, truer, purer soil. In later permutations of this perspective among environmentalists, human beings themselves, with all the qualities that seem to put us outside a reductive view of "nature," would appear to be a monstrous threat to the natural order.

In a similar vein, American transcendentalism found its faith, hope, love, and redemption in wilderness. Thoreau famously remarked, "In wildness is the preservation of the world." After reading Emerson, Muir once invited the great transcendentalist, whom he had met in 1871, to join him "in a month's worship with Nature in the high temples of the great Sierra crown beyond our holy Yosemite."[15] (Emerson declined.) Yosemite may not be Chartres, but it is difficult to deny that in all these exalted feelings about nature, something like the passion for the beauty and meaning of the Virgin had assumed a characteristically American form, which had analogues in other parts of the world touched by the Romantics.

The World Is Too Much with Us

Yet except for the minority Muir represented, Adams was right that a profound ambivalence reigned over notions of natural value in the "real world" of American daily life. The Dynamo was highly successful and — in Adams's view — often highly vulgar in its unholy alliance with commerce. In this, he, like Muir, sounds a characteristically modern intellectual note that continues to reverberate inside several forms of environmentalism (free-market environmentalism, one of the more exotic and promising blooms on the American ecological landscape, had not yet made its appearance).[16]

Piety and practicality, as the great historian must have known, had coexisted in the past. Chartres Cathedral itself was the product both of devotion to the Virgin and of the most advanced medieval technologies. The surplus wealth dedicated to building it came from expanding commerce. In his beloved Middle Ages, the cultural and religious forces of the Virgin had emerged from and shaped a society that had its own large share of vulgarity and crude commerce. But this appeared to Adams to have no relevance in the modern context. In some ways, his attitude was understandable because, at the time, the Dynamo had put unprecedented power in the hands of businessmen. And in many places, industry had attacked nature as never before. As usual, it was John Muir who put the case most forcefully: "The temple destroyers, devotees of ravaging commercialism, seem to have a perfect contempt for Nature, and instead of lifting their eyes to the God of the Mountains, lift them to the Almighty Dollar."[17]

Whatever the ambiguities of emerging industry, many of the current features of the environmental movement had fallen into place by the beginning of this century: the romantic opposition of nature to civilization, casting everything that was characteristically human as almost outside the natural order; a particular animus towards commerce, without recognition that increased human power and wealth made human life and a more contemplative view of nature easier; the exalting of a religion of immediate nature over age-old spiritual traditions; and a kind of elite cast to it all (Adams and Muir did not, like the common folk, still have to scratch out a living from the land or from industry; those who did were not likely to think their labors mere idolatry of the Almighty Dollar). Among those who had become aware of the new environmental problem, a pair of stark alternatives had arisen: Either nature should continue to be used for human good, or it should be preserved unspoiled. For Muir, even the "wise use" advocated by Gifford Pinchot, the first director of the U.S. Forest Service, represented a kind of sacrilege. In the modern environmental debate, those who would permit use and those who advocate wilderness preservation have become virtual warring denominations.

A More Catholic View

The following pages, as will become immediately evident, are written from a specific point of view. As a Catholic and an American, I take a far different approach to questions of the Virgin and the Dynamo than any of the previously mentioned figures. To begin with, in Catholic theology not only are man and nature part of the same cosmos, but so are human psychology, social systems, politics, science, industry, commerce, and inventiveness. In this view, there can be no romantic retreat from complex human questions to an oversimplified nature, because it is nature and nature's God that have made us more complex—for good and ill—than other creatures. Indeed, the flight from full humanity is one of the characteristic marks of a wrong understanding of man and nature. All of us need periodically to simplify our lives, in a different sense, to discover or rediscover our particular vocation in the world. Whole societies from time to time require a return to essentials. But no spiritual practice or social program can remain true to the fullness of *human* nature, as God and nature have made us, if it does not recognize that human creativity is an image of God's creativity, with the special prerogatives and responsibilities entailed by that status.

Consequently, the following pages do not fit easily into the usual dichotomy: environmentalism or developmentalism. On the whole, I believe science and technology have benefited us in ways for which we should be deeply thankful, and they can continue to do so in more environmentally benign ways in the future. Without blaspheming, it can be said that today the hungry are fed, captives freed, the lame walk, the blind see because human beings have made systematic efforts to understand and utilize nature and to distribute goods through efficient economic arrangements. There are sometimes very high prices to be paid for such benefits. People who think development has been on balance a real human good cannot simply dismiss every environmental worry as masking totalitarian impulses or a hatred of technology. If the notion of original sin means anything, it is that, even as we advance in knowledge and power, evils too will multiply. The great modern theologian Hans Urs von Balthasar observes that Jesus said the wheat and the tares *grow together*.[18] So we have to be appreciative of, and grateful for, the great growth in human invention, industry, and commerce, even as we are critical of some of the consequences.

A similar mix of appreciation and wariness is necessary towards the usual forms of environmentalism, both religious and secular. Environmentalism at its best has alerted developed societies to questions that desperately needed asking. The result has been an enormous improvement in environmental quality and in our very thinking about how to operate in a global context where human activities now have far-reaching consequences. The claims of environmental organizations and their critics notwithstanding, if you allow for human imperfection and shortsightedness, environmental efforts have been largely a success, though the media rarely report as much. Air, water, and soil are in better shape today in many parts of the world than they were twenty-five years ago. And greater improvements lie just around the corner.

Yet wariness towards the environmental movement is still justified. The same Bible parable that describes the wheat and the tares growing together also warns against the simplistic belief that we know how to pull out the weeds without destroying the grain. This is no recipe for merely accepting pain as the price of progress, but a call for moral realism and a warning that we can lose the staff of life if we are too sure of our own ability to discern evil. Many environmentalists give the impression that their concerns must trump all other considerations. But in these matters, as in others that involve the complex concerns of the human, we need to learn how to live in various tensions without either acquiescing in gross exploitation of nature or expecting impossible purities not given to us in this life. The environmental movement has raised questions that were sometimes overlooked or even unknown prior to recent decades. But it has not yet proven itself to possess the steady practical or theoretical wisdom that will put bread on the table for us or the billions of poor on the planet.

The biblical tradition with its moral realism might offer good guidance here. To be sure, we cannot simply open the Bible or the books of some celebrated theologian for answers on such matters as auto emissions levels or wetland preservation. That is not how a good reading of the Bible or of theology approaches authentically new situations. Theological thought develops as questions are put to it. Notions about God, Jesus Christ, morals, and church authority were elaborated as people tried to bring the full light of revealed truth to bear on specific disputes.[19] We have only begun to think about what revelation tells us about environmental questions. No religious figure of the past, not even the nature-friendly Saint Francis of Assisi, answered "environmental" questions, because they did not yet exist in any significant sense. Earlier ages may have used natural resources unwisely, but it was only when the human race vastly developed its powers in recent centuries that environmental issues, as something more than a passing disturbance in nature, could arise.

Preparing the Next Phase

We are all only tentatively beginning to figure out what to do with the immense power we have achieved over the natural world. Some situations call for prophetic denunciations: individuals and businesses do foul water and air out of sheer greed and indifference. Law and conscience will have to deal with them. Others are asked to balance real human needs, and sometimes sheer economic survival, against environmental concerns. And much human activity — even industrial production — is far more environmentally neutral than we typically acknowledge today. You do not have to be an interested party in one of these disputes to think that we need a much more articulated approach to the environment that will distinguish real crises from necessary trade-offs, tolerable environmental impacts from the intolerable, and special cases from universal rules.

The following pages, therefore, advocate intelligent development. I deliberately do not use the common term "sustainable development" because that concept, as we shall see in later chapters, has been laden with many highly dubious assumptions. To anticipate briefly, sustainable development generally makes the socialist mistake of thinking that a central bureaucracy can plan for the operation of a whole economic order better than the innovators and entrepreneurs within it. The latter groups are generally vilified in the environmental literature, but we may rapidly be coming to the recognition that, without the kinds of competition that will lead to cheap alternative energy sources, still greater agricultural yields, and more efficient operations in every field, our environmental situation would be dire indeed. By contrast, the environmental planners usually take quite pedestrian approaches to current problems and often interpret the concept of "sustainable development" so strictly that they inhibit the necessary transitional phases of "unsustainability" that eventually lead to better conditions.

Benefits from Markets and Industry?

Furthermore, religious people who become exercised about the environment have an ingrained tendency to view technology and industry as by nature greedy and exploitative — and not only in their effects on nature. Mainstream religious bodies and some not-so-mainstream ones have had trouble accommodating the self-interested pursuit of profit that is the engine of general prosperity in market societies. The profit motive can be a serious moral evil, but it can also be quite benign. As that devoted Christian Dr. Johnson once said, "A man is rarely so innocently occupied as when he is making money." Johnson understood greed, but he also understood that much of commerce and industry is a neutral matter. Viewing industry and corporations as by nature inclined to exploit human and nonhuman nature evokes powerful religious images that, in the vast majority of cases, I believe, are simply mistaken. Indeed, if we believe that our God-given powers reflect a Providential design intended to be used both for our good and for the good of the planet, we might take a quite different view of the freedom to respond to human need and natural problems that markets alone make possible.

The average environmentalist in a developed country partakes of some of the same blindness toward nature that he imputes to the greedy capitalist. Food and the other necessities of life are readily available. Except when an occasional storm blows through or a river overflows, nature seems no real threat. If we read history carefully or even look at the lives of most people in underdeveloped countries, however, we see that our situation is privileged indeed. Most people at most times in history died early from "natural" causes. Wild animals and natural catastrophes were only the most lurid parts of the story. Famines, epidemics, and chronic overwork in quest for subsistence were the more usual daily threats. Paradoxically, it is our very success in meeting human needs that makes concern for nature humanly feasible.

Unfortunately, environmentalists tend to have one volume level for every problem. Religious environmentalists, too, seem to regard everything from an accidental oil spill to a polluting smokestack to topsoil erosion as all part of one interlocking apocalypse. I take a different view. A few issues are clearly pressing and may have large-scale effects on what we must do to be responsible stewards of creation. For example, the steady buildup of greenhouse gases in the atmosphere demands attention. The science here is uncertain. There is no clear evidence at present of human-caused global warming, and it seems that much of the buildup preceded the increase in human emissions of greenhouse gases. Yet can these levels be allowed to rise indefinitely without consequences? Without working ourselves into a premature panic, we want to make sober and careful assessments of what consequences, if any, may follow and what steps, if any, we can take. Mere action without solid knowledge of what we are doing may make us feel good but, like the efforts of the Buddhist animal rescuers, may unintentionally do harm.

By contrast, issues such as air and water pollution and problems caused on land by chemical fertilizers and pesticides are, it seems to me, largely on their way to solution. The Green Revolution, together with new inventions such as shallow- or no-plough techniques and novel irrigation methods, can provide enough food for as many people as are likely to be alive when the population stabilizes at around ten billion on *less* land at no significant environmental cost. Because of new farming techniques, there are already more undisturbed forests and animal habitats in America and Europe than there were a century ago, and the next century, contrary to dire predictions, will, through technological advance and human prudence, do even better. People living in cities experiencing suburban sprawl, which is to say most of us today, may not believe it, but it is precisely because more people are living in cities and intensive farming techniques have reduced the need for people to work the countryside that lower human impact on the land is possible.

Population pressures, in my view, have been grossly exaggerated, and there is good evidence that population will stabilize at quite manageable levels as development spreads around the globe. The antihuman pronouncements of Paul Ehrlich, Garrett Hardin, the Rockefeller Foundation, and a host of other individuals and institutions reveal a real and open callousness about human beings. A few have even taken the old Malthusian line that it is wrong to feed people in poor countries with growing populations because of the impact it will have on the earth. Curiously, many religious figures also seem quite willing to take this antihuman line and promote abortion, sterilization, and aggressive contraceptive programs while relatively neglecting measures that might provide people with the necessities of life.

Only one thing makes population a threat in current circumstances: political and economic chaos. Anyone who has ever seen, for example, the border between Haiti and the Dominican Republic has had graphic proof of how politics affects the environment. Haiti and the Dominican Republic share the island of Hispaniola, where Columbus set up his first major settlement in the New World. But their histories over the years diverged. Haiti came under French rule and then became independent. Its corrupt and incapable governors left it in poverty and constant crisis. The Dominican Republic, too, has had its share of corruption and poverty, but with less instability. Today, at the border, the Haitian side presents mountains stripped of forests and eroded down to bare rock, the Dominican side, lush forest — graphic testimony to the role even modest political and economic stability may have in preserving the environment.

Environmentalists are right to remind us of the spiritual narrowness and potential cataclysms we face as a result of our misuse of nature. They are wrong in thinking they can simply subordinate the complex *human* world of religion, science, philosophy, art, politics, and economics to the physical environment, conceived of as a stable standard. Nature has the potential to speak to us spiritually; we need to recover wisdom by contemplating both what is present and — a point often forgotten — what is absent in nature. Nature also speaks to us practically. In the recent past, we have paid attention primarily to the most immediately useful lessons to be learned from nature. But even self-interest now will lead us to recognize that there are larger-scale side-effects to be included in gauging our activities. Finding a way to make room in the world for the quintessentially human things while respecting and safeguarding the environment is what ecology, the science of the household, must be about.

The Structure of This Book

This book will proceed along two large lines. In Part One, I will try to establish what the wider tradition of religious and philosophical thought in the West might have to say to us about environmental issues as those are illuminated by science. In Part Two, I will examine how several prominent religious figures in the current debate are advancing or retarding wisdom about man and nature. For the most part, I find people's principles on environmental issues more defensible than their proposals for action. As far as possible, I want to address their thought, although it is impossible not to take into account the translation into practice.

For those unfamiliar with the debate over Western civilization's legacy on the environment, it is worth remarking that the question is more difficult, complicated, and controversial than may first appear. To begin with, it has become common among people deeply dissatisfied with developed societies and economies to see "Western philosophy since Plato" as simply irredeemable. Postmodernists, feminists, environmentalists, and radical social critics disagree about many things, but they seem united in the facile dismissal of Western thought since the Greeks. The British philosopher and mathematician Alfred North Whitehead once described all Western philosophy as a "footnote to Plato." At the time, it was meant as a compliment to Plato and the West. Since then, most undergraduates and even elementary and high school students have been encouraged to believe that a large part of what is wrong with the modern world began with the secular thought we inherited from ancient Greece. Plato and Socrates are also often taken to task for having turned Western thinking toward "linear" rational discourse and away from the more intuitive, organic reflection on nature that allegedly exists in other traditions. In short, many students today are dangerously taught to see one of the great foundations of civilization as merely an unnatural imposition of reason over feeling.[20]

The picture is not much better when we turn to environmental views of the Western religions — Judaism and the various forms of Christianity — that originate in the Bible. The command "be fruitful and multiply and have dominion over the earth" in the first chapter of Genesis is often taken by environmental critics to be another source of Western haughtiness toward the natural world. Allegedly, from that source flow Western science and technology, along with unrestrained notions of progress and consumption. There may be a kernel of truth in this charge, but, as will soon become apparent we cannot attach much importance to such a simple assertion when many other factors have operated throughout history. The human race, at the time of the writing of Genesis and for centuries thereafter, faced a fierce foe in nature. And all premodern cultures, not only those shaped by the Bible, felt few qualms about restraining natural threats or fulfilling their own needs, even though they may have engaged in elaborate religious rites before acting to avoid angering the gods or making them jealous. But it is only in the last century or two, when mankind has won many battles against threats *from* the environment, that humans have had the luxury and leisure to begin contemplating what a broader stewardship of the world requires. In that longer perspective, our technological prowess is nothing to be ashamed of and may even be thought of as providential. Indeed, *pace* Thoreau, in civilization may be the preservation of the world.

The Copernican Revolution in Blame

Science and technology are Western creations that owe something to the Greeks and the Bible. But blaming Western philosophy and religion for science and technology represents a sharp change in historical analysis. Not long ago, progressives in general and scientists in particular were criticizing ancient philosophy and religion for being the *enemy* of scientific enlightenment (recall T. H. Huxley and the strangled theological serpents). A whole school of historians of science has shown convincingly that science emerged in the West because of several factors, not least biblical notions of creation. In the Bible, creation is distinct from, though not wholly unrelated to, its creator. Only when a religious system allows for secondary causes in a somewhat self-subsisting pattern can nature in the modern sense exist. Further, the separation of Caesar and Christ, of church and state, initiated by the New Testament and publicly emerging in the Middle Ages, gave researchers and universities an institutional independence that they did not have outside Western Christianity, not even in the Christian Orthodox East.[21] Western religion and science, then, are not as opposed as the proponents of either may have assumed.[22] But is this reason for praise or blame?

Obviously, given the paradoxical nature of much of human life, the answer is, both. The nub of the problem is the vast difference between science as it emerged from Western beliefs — when it still looked upon nature as a divine manifestation — and the limitless technology that developed when religious influence was waning in Western societies owing to the impact of an overly scientific approach to nature. It is no accident that Francis Bacon, writing in the seventeenth century and advocating the separation of religion and metaphysics on the one hand from scientific investigation on the other, speaks, in violent terms, of "putting nature on the rack to answer our questions." But Bacon places this phrase under the aegis of quite noble purposes: "For the glory of the creator and the relief of men's estate."[23] We may think that Bacon and his successors went too far. We may forget, however, about the widespread miseries even as late as the Elizabethan age, which Bacon helped to banish from human life.

Much time and effort have been wasted in trying to determine whether the Bible is pro- or anti-nature in the environmental sense. On the negative side, there is the passage about human dominion, though another way of looking at the relationship is that proper dominion is something nature seeks. On the positive side, a whole series of texts could easily be assembled from Genesis, the Psalms, Job, and the Wisdom books showing the goodness of creation, God's glory revealed in natural processes, and the mysteries beyond human understanding that creation contains. The biblical tradition generally thinks of nature and revelation as two Divine Books correlated with each other. As the Eastern Orthodox theologian Kallistos Ware has observed, Christians try to discern "in and through each created reality, the divine presence that is within and at the same time beyond it . . . to treat each thing as a sacrament, to view the whole of nature as God's book."[24] But to read the book of nature properly and fully, both careful scientific attention and divine revelation are indispensable guides.

The Inescapable Human Difference

For the environmentalist who worries that man's dominion over other creatures found in the Bible vitiates all the other praises of nature, I see no other response than to say that, whatever misuse we have made of such dominion, that's simply how it is. Premodern non-biblical cultures, lacking our many modern weapons against a nature "red in tooth and claw," certainly had their own utilitarian attitudes toward nature. Even today, real Native Americans, as opposed to the romanticized nature-lovers we often read about, seem to have very little sentimentality about nature. Anthropologist Åke Hultkrantz has argued that

> we seek in vain for a pure aesthetic beholding of nature [among Indians]. An educated Senecan Indian well initiated into the traditions and values of his people once told me of the widely divergent reactions of Indians and whites when they move to new regions. The white man, he says, is impressed by the appearance of the landscape and meditates over its beauty, whereas the Indian first of all asks, where are my medicines?[25]

Indians in day-to-day contact with nature know what a stern sister she can be. Nor can they be looked upon, by and large, as somehow not showing the evil impulses imputed to the Bible and the West. About twelve thousand years ago when the last Ice Age retreated, for example, inhabitants of North America seem to have contributed to the extinction on this continent of woolly mammoths, saber-toothed tigers, giant sloths, and horses (the horse only returned when Europeans brought it back during the Age of Discovery). More recently, we have learned that the tribes Europeans first encountered in what they regarded as untouched wilderness had vast influence on their surroundings.[26] What we think of as the natural condition of these shores was already the product of human intervention in nature.

This book is in many ways a sequel to my earlier volume *1492 and All That: Political Manipulations of History.*[27] In writing that work, I tried to argue that the current tendency to see the expansion of European civilization over the globe as a kind of imperial infestation is badly one-sided. Europeans perpetrated evils on a worldwide scale, and societies of European origin continue to do so within their borders. But that is merely to say that they are human entities with all the grandeur and tragedy of the human spirit. My judgment is that much of the current criticism of Western religion, politics, corporations, and societies stems from a profound misunderstanding of the historical record and a demonic drive to discredit many of the very features of Western societies that make current critiques possible.

Criticism of Western impact on the environment, in particular, generally fails to take into account the very Western notions on which such criticism is based. However unpopular it may be in many modern religious and nonreligious schools of thought, the beginning of wisdom in environmental matters, absent some unlikely widespread divine illumination, is the recognition that man is a rational animal—that is, an animal with an important difference. Even if it were possible for a person to live like an animal in nature, a hypothetical case that seems very unlikely, few of us would choose such a life. We can live closer to nature, we can refuse to cause gross harm to the world in which we live, but ultimately we have concerns that nature does not understand, and we have to make some choices about our inevitable use of resources for activities that are uniquely and inescapably human.

To say that we are rational animals means, in the classic tradition, that we have some powers of foresight and control over our lives. This ancient view of rationality should never be confused with more modern conceptions, particularly since Descartes, that sought to give a narrow rationality total and transparent mastery over self and nature (though properly contextualized, Descartes' discoveries still have many uses). In that old modern view, nature became a vast machine that, with enough effort and patient observation, could be turned to human purposes. Those purposes and the human person were somehow ejected from "nature" so conceived. Curiously, many environmentalists, fleeing the nightmare of scientific totalitarianism that they believe brought us to our current crisis, go on to embrace the very premises of their technological adversaries: they believe human knowledge somehow makes it possible to produce absolutely balanced and sustainable relations with our environment. Many scientists are wary of such pretensions; but the strong impulse toward comforting fantasies and pseudo-religious environmental myths makes this view particularly difficult to dissolve.

Even when environmentalists concede the complexity of the task, they seem to converge on a vision of human life that is long on reducing expectations and on fear of the future, short on hope or imaginative responses to problems. The typical diagnosis and prescription runs something like this:

One of the most crucial house rules we must learn is that we are
not lords *over* the planet, but products *of* its processes; in fact, we
are the product of a fifteen-billion-year history of the universe and
a four-billion-year history of our earth. We are an intimate and in-
tegral part of what we want to know; planetary knowledge is self
knowledge. Hence, ecological knowledge is not about something
"out there"; it is about ourselves and how we fit into the scheme of
things. The most important ecological knowledge we can have,
then, is not how we can change the environment to suit us (a tactic
that may, in the long run, be not only impossible but disastrous for
ourselves and other species), but rather how we can adjust our desires and needs to what appear to be the house rules.[28]

There is much good as well as bad jumbled in this brief description,
and this book is an effort to separate the wheat from the chaff in such
characterizations.

The Varieties of Ecotheology

Most environmentalists want to be able to praise the cosmological
process and God's providence over it, including the development of
human beings, down to some point at which things supposedly took
a wrong turn. For cultural historians, as we shall see, that point may
come after an idealized neolithic period, when we began to exploit
land through agriculture and build cities; for feminists, it may come
at the rise of the great civilizations with an alleged transition from
worship of Mother Goddesses and female dominance in societies to
the sky and warrior gods of "patriarchal" cultures; for social critics,
the rise of capitalism, Europe, and the nation-state marks the evil
moment. Each of these analyses is, I believe, wrong and a dangerous
misreading.

Many ecotheologians desire a return to some idealized past, quite
often the primitivism of drums and tribal existence. What we are
then to do with the undeniable human glories of Mozart or the
splendors of city life at its best, they cannot say. If human beings are
part of the 15-billion-year development that modern science has recognized, we need to find a way of living in the world that accommo-
dates *our* cosmic development and nature as well as the simple and
common bases of all life.

Human creativity and the shaping of nature are both part of the very cosmic process unfolding. Our lives, like some of the phenomena discovered by quantum physics, show elements of creativity and spontaneity that are not simply reducible to machinery. As the physicist and Anglican priest John Polkinghorne has said of our current knowledge:

> The future is genuinely new, not just a rearrangement of what was there in the past. In such a world of true becoming, with its open future, we can begin to understand our own powers to act and bring things about. Such a physical world is capable also of being open to God's providential interaction and his agency. Our whole picture of the physical world is much more hospitable to the presence of both humanity and divine providence than would have seemed conceivable a hundred years ago.[29]

It is impossible for human beings to find their place in nature as that seems to be conceived in most environmental literature. The cosmic process shows us many things. One of them is that the emergence of higher levels of organization in the story of the universe establishes levels of meaning and complexity — hierarchy, to use a much decried term — that cannot be explained, limited, or directed by lower levels alone. The first single-celled organism's needs to secure nutrients and expel wastes were more complex than anything found in any inorganic system. Inorganic processes contribute to the simplest form of life; they cannot be a complete guide for it, because life processes exceed mere physical processes.

A similar, but far more momentous, shift occurred with the appearance of human beings. It is a question for philosophy and theology whether human beings are "higher" organisms than others and what, if true, that might mean. At an operational level, however, we can say that human beings have both freedoms and responsibilities with regard to the rest of nature that no other being known to us possesses. We do not say, for instance, that one primitive organism selfishly almost killed off other living beings on the planet by its unrestrained reproduction and emission of the then highly poisonous waste gas we call oxygen. We regard what happened as a natural process in which an organism and an environment interacted in certain ways that, it appears, could not have been any different. Our human reproduction and emission of industrial gases partake of a wholly different realm, of knowledge and choice, of moral decision and worship.

The usual ecological appeal to human beings is to remind us of our relationship to the rest of nature. That is all to the good as a corrective. But we need to deal with the fact that we are also quite different from the rest of nature. As the late American novelist Walker Percy said, a conscious being who uses language does not exist in an environment but in a world.[30] Most science of the past few centuries dismisses this distinction as literally meaningless. The crude facts of a mechanized nature are the reality; the world we all see around us is a mere product of the interaction of the natural mechanism and our senses, a product that is then assembled into a "socially constructed" picture of reality. Hence the various medical and psychological attempts to harmonize our thoughts, emotions, and actions with "nature," not in Plato's transcendent religious/scientific sense, but in the narrower sense of environment. Percy, who was a psychiatrist as well as a novelist, knew where shearing off the specifically human things would lead us: "Show me a man trying to live like an organism in an environment and I'll show you one of the most screwed-up creatures on earth."

The same might be said of entire societies. The human species is not just one more among others; for us to treat ourselves that way will do neither us nor nature any lasting good. It is only by our recognition of our own unique place in nature, which, contra many of the most prominent ecologists, puts us both within and outside the web of physical forces, that the properly human may flourish without becoming a threat to other species. Without the Virgin as the bedrock of our actions, the rapaciousness of the Dynamo will inevitably be the pattern for human society. But without the full religious appreciation of what the Virgin entails, our efforts to tame the Dynamo can only result in a weaker form of mechanism or an unsatisfactory pantheism — exactly what we see in most "religious" environmentalism — that will get both the truth and the value of man and nature dangerously wrong.

In Better Hands

The relative emphasis on meeting material human needs in recent centuries has caused much environmental damage and portends even more. But the race has come to see the consequences of its acts, is taking steps to reduce the most serious of them, and is not facing an apocalyptic threat because of its own successes anytime soon, or perhaps anytime at all. The earth inhabited by humans will never again look exactly like the earth before humans or at any time in the past. But part of the religious wisdom we need to bring to bear in such a situation is precisely the virtue of hope — hope that we will, *Deo volente,* survive into the future and even continue to flourish in the complicated ways that a rational animal prone to spectacular evils and goods may flourish.

What, then, is to be done? If the old scientific view of man as just another animal holds little promise for transcending our limitations, and if pseudomystical and romantic notions of nature do not reflect reality, is there any path left for us? It is a central contention of this book that an answer to some environmental questions may still be found in the classical religious views of the West, supplemented by science and the wisdom of other traditions—that is to say, in the recognition that both the Virgin and the Dynamo are necessary to a full human life and the ongoing evolution of the universe. We may treat "environmental" problems only if we first admit that we are the kinds of beings that live, not in an environment, but in a world. Indeed, the Talmud warns that to destroy a single person is to destroy a whole world. And in the world revealed to us by the Bible, we can only speak of solving *some* of the environmental questions, for the greatest contribution the biblical tradition, properly understood, might make to our current predicament is to remind us that we do not have all the answers — about anything. We are not the Creator, and the best we can hope for in a world that by its very nature exceeds our mastery is a reasonable hope about the human future.

The drive toward mastery, as many modern thinkers have pointed out, was the root of the problem in the first place. For an unbeliever, the lack of any discernible order in the old universe of mechanism may point toward nihilism or the postmodern "abyss." For a believer, it is a confirmation that we are not, and can never be, complete masters of our own destiny. Our intuition that something good and worthy of love exists in the universe, over and beyond the positive and negative dimensions of nature as we experience them in this world, should lead us to see that there can be no finally complete science, of

PART ONE

On Deep Background

1

The Bible Made Me Do It?
Creation Lost, Found, Mislaid

In the Popol Vuh, the Quiche Maya creation account, there is a curious story about how the gods make several abortive attempts to bring into being a special kind of creature. Some of their first tries are made from wood, but those beings are too rigid; other figures are formed out of clay, but they wash away in the rain; still others turn out to be animals. Seeing them, the gods lament: What kinds of creations are these? They can only grunt and twitter. We need beings that can speak, understand the motions of the stars, and worship us. So the gods tell the animals: Be content; you're going to have to live out in the forest and canyons. "Just accept your service, just let your flesh be eaten."[1]

That kind of natural hierarchy, in which human beings are special creations because of their capacity to understand the world and the gods, and other beings are subordinate, appears almost universally, with local variations, in non-Western creation stories as well as in the Bible. Despite the worries of many in the environmentalist movement about human hubris vis-à-vis nature, there is no getting around the widespread testimony to special human status. One currently popular counterstrategy is "biocentrism," in which all living things and the habitats they require are to be placed at the center of our thinking about nature (a concept that certain religious people confuse with reverence for the creation). But this strategy conceals the fact that biocentrism itself is a human stance. Various currents in contemporary environmental thought abhor the very idea that we get to decide how to look at nature. For them, nature has to have independent, controlling status over us; otherwise it only continues in humiliated form on human sufferance and will always be vulnerable to human depredations. That was the argument, eloquently presented but ultimately self-contradictory, of Bill McKibben's highly influential book, *The End of Nature*. So strong is the impulse to put nature above human beings that the biologist Edward O. Wilson, who seems absolutely delighted with all other results of evolution, remarks, "It was a misfortune for the living world in particular, many scientists believe, that a carnivorous primate and not some more benign form of animal made the breakthrough to intelligent control of Earth."[2] But who are these scientists, if not "carnivorous primates"?

Whether we look to scientists or religious thinkers in the environmental movement, there is a strong tendency to portray the human race as just one more kind of animal, part of a vast, ongoing cosmological process that began fifteen billion years ago and continues to bring forth new and valuable beings and events. But so far as we know, human beings are unlike all the other products of that cosmogenesis, in that to a certain extent we understand our place in that process. We alone are curious and capable enough to look back across that vast expanse of space and time and weigh its significance. We alone do not merely fall asleep or play when our physical needs are satisfied. We alone probe the meaning of who and what we are and what our prerogatives and duties may be toward the inanimate objects and other living beings in the world that will never understand how we treat them.

Power and Responsibility

The Bible taught us quite early about the immense responsibility as well as power put into human hands. But since Lynn White's famous 1967 essay, "The Historical Roots of Our Ecologic Crisis," it has become common to blame the monotheism of the Bible and the special place it accords human beings for what White called "the most anthropocentric religion the world has seen" and for the scientific exploitation of nature.[3] The political philosopher Ernest Fortin has described this as a "the Bible made me do it" defense.[4] White's thesis has been challenged or qualified in various ways in the past thirty years. In particular, critics have pointed out that it is difficult to ascribe the rise of a complex entity like modern science and technology to a single historical cause like the Bible, and it would be even more difficult to explain how the effect has grown stronger as its cause has grown weaker in recent centuries. But even though the Bible and Popol Vuh and any number of other creation accounts may share similar notions of hierarchy and human uniqueness, in the environmental literature it is usually only biblical religion that is thought to sow the dangerous seeds of anthropocentrism.

Whatever ecological problems the Western religions may have contributed to, however, it is worth noting that the Bible never gives anyone unqualified power—no priest, no prophet, no king—outside the divine law and love for all creation. In fact, it would be truer to say that by biblical standards, the greater the power someone possesses, the more severely its use will be judged.

There is no denying that from the outset, the Bible shows God giving humans a full portfolio. He tells Adam and Eve, "Be fruitful and multiply, and fill the earth and subdue it; and have dominion over the fish of the sea and over the birds of the air and over every living thing that moves upon the earth" (Gen. 1:28).[5] This comes even before the Fall and the disorders within and outside us that followed. Among many similar biblical passages, perhaps the most beautiful in its expression of surprise at our cosmic status is in Psalm 8:

> When I see the heavens, the work of your hands,
> the moon and the stars which you have arranged,
> What is man that you should keep him in mind,
> mortal man that you care for him?
> Yet you have made him little less than a god;
> with glory and honor you crowned him,
> gave him power over the works of your hands,
> put all things under his feet.
>
> All of them, sheep and cattle,
> yes, even the savage beasts,
> birds of the air, and fish
> that make their way through the water.[6]

In the Sermon on the Mount, Jesus told his followers to consider the birds of the air, whom God feeds, and not to be anxious, asking, "Are not you of more value than they?" (Matt. 6:26). In sum, the Bible asserts both a hierarchy, with humans at the top among the earthly creatures (though not the heavenly), and the greater value of human beings.

For many environmentalists, these two notions are the root of all evil, the original eco-sin. When Gregg Easterbrook, a liberal and an environmentalist, mentioned in passing during a lecture at the Harvard Divinity School in the early 1990s that "people are more important than plants and animals," he was hissed.[7] But if, with some theorists of nature, we believe all beings are on an equal plane, we immediately encounter absurdities. The AIDS virus has as much "right" to fulfill its own needs as the AIDS patient. The dust mites in your carpet have as much right to be in your home as you do to walk across the living room. Some theorists have recently argued for the rights of trees and even of rivers and rocks.[8] Anti-hierarchy notions are deeply seated at universities and in our cultural life as a whole, even among our religious thinkers.

What Went Wrong?

The power and status we have from God is always linked with responsibility in the Bible, but if that is so, why did the West, with its biblical heritage, come to find itself with environmental problems? Perhaps the first thing to say about this question is that we forget it was the West that first recognized the problem and its own obligations. No one answers a question until it has been asked. All human societies, from Australian aborigines to Mayans to modern Americans, make use of natural resources for their own survival. Only relatively recently has human power grown to the point where it required the kind of global ecological analysis we need to undertake today.

So environmentalism, too, is necessarily a recent phenomenon, more politicized and ideological than its more scientific cousin: ecology. The word ecology only appeared in the middle of the nineteenth century when the German biologist Ernst Haeckel coined the term to denote the scientific study of "nature's household." As sciences of a sort, eco-logy and eco-nomics are related. Each of them takes the household or world, the *oikos,* as a starting point. In economics, as Aristotle first observed, we try to understand what we need to live a good life in our households. In ecology, as we now understand the term, we do the same for the larger household we call the earth.

Environmentalism is the moral and political debate over how to reconcile these two households. Science and religion shape, but cannot replace, that debate. During the 1997 Kyoto conference on global warming, for example, we heard that in a decade scientists will know whether anthropogenic global warming is real. Perhaps so. But even if we find that we are responsible for climate changes, as with all political questions, we will still be left with the decisions of what to do about it. And depending on what we value most, we will choose paths that favor either certain kinds of development that we hope will be good for us and the planet or less human activity or some combination of the two — exactly the alternatives that exist today.

Physics and Metaphysics

Metaphysical notions shape our vision of the world, often covertly, so that what may seem abstruse questions of the theology of human and nonhuman nature have become quite important. Most Americans form their moral judgments on religious grounds. The late Carl Sagan, no friend to religion of any kind, was instrumental in setting up a joint scientists-believers environmental organization, he admitted, because religious passion was the only force likely to mobilize Americans for the environmental work he thought needed to be done. Similarly, the biologist and champion of biodiversity Edward O. Wilson, a member of that organization and also a nonbeliever, has argued in religious language that "biodiversity is the Creation."[9]

Some might regard this use of religion as cynical, but we owe a debt to all those who have raised questions about the human impact on nature, scientists and environmentalists, believers or not, because they have called our attention to a new and serious human challenge to which we have begun to rise. And one of the ways to repay the debt is to separate the good science and moral passion from the errors in larger perspective that often accompany them. Despite the harm it has done to the environment, only the Western tradition, with its two cultures, its Virgin and Dynamo, is in a position both to defend the full truth of man's uniqueness and to make judgments about what is really good for nature. For our present purposes, it is worth noting that Western religion is one of the sources of science and technology, and one of the restraints on their unbridled development. The doctrine of Creation implies that the physical world is a willed order that we may partly understand because, though divine in origin, nature is not divine in itself. Biblical religion may be somewhat anthropocentric in a weak sense, but it is far more profoundly and suggestively theocentric. The combination of these elements creates tensions and complexities that hold both great promise and peril.

That fact was not lost on some of the founders of modern mechanical views of nature—or their critics. In the seventeenth century, Descartes, one of the bugbears of environmental thinkers, repudiated the earlier Christian philosophy of nature that was based on both Aristotle and sacred Scripture because he thought it was not useful:

> instead of that speculative philosophy which is taught in the schools, we may find a practical philosophy by means of which, knowing the force and the action of fire, water, air, the stars, heavens and all other bodies that environ us, as distinctly as we know the different crafts of our artisans, we can in the same way employ them in all those uses to which they are adapted, and thus render ourselves the masters and possessors of nature.[10]

It is common to point out the hubris in the concluding phrase, but few recognize Descartes's aim of remedying real human suffering and misery, whatever the other effects on a culture that came to ride that particular truth too exclusively.

The uncertainty about where authentic concern ends and hubris begins draws our attention to the need to get clear in our minds the basic biblical perspective behind this double phenomenon. Among the many contradictions of our current situation, we seem to believe, on the one hand, that nature is a sacred realm that must be preserved and, on the other, that nature is merely a collection of energy and matter in constant flux. To acknowledge this, however, immediately alters the simple notion that the order of nature as we find it at any moment is eternally in balance and sacred per se. Nature in its totality may be a divine gift and its dynamic order an inspiration as to how we should live, but we need to look elsewhere for the sacred and for stable guidance as to our actions.

Religious thinking about nature in the West has a complex history. In addition to the varied influences of the Bible — a library of books rather than a single volume, whose writing extended over several thousand years — several other strands have been woven into our thinking. Ancient Greece, with its philosophical brilliance and rudimentary, tragically truncated mathematics and science, recurs periodically in Western history as a new spur to reflection. Diverse theological and philosophical currents associated with great figures like Augustine, Aquinas, Luther, Calvin, Leibniz, Spinoza, Descartes, Kant, Hegel, and many others further complicate the picture — and these are only the Western thinkers within the Judeo-Christian horizon, broadly construed.

These influences have been traced in various intellectual histories, and it would be beside our main point to go into them in any depth here.[11] Furthermore, our current situation, marked as it is by unprecedented ecological concerns, leaves much previous speculation without immediate purchase, however important it may remain in the overall realm of Christian and Jewish thought. For a first approach to the questions of the Virgin and Dynamo in our age, we need a more immediately serviceable map of a wide range of problems from immediate environmental issues to the deep roots of our sense of nature. Let us turn to two great Christian expositors, one ancient and one modern.

A North African Theologian

The first great and widely influential Christian theologian in the West was Saint Augustine (354-430), bishop of Hippo in North Africa. Augustine lived a wild life as a young man and repented of it in his maturity. He is often accused today by religious environmentalists of being overly moralistic, puritanical, antinature, antisex, antifun, misogynistic, excessively inward, individualistic, and otherworldly. Anyone who cares to examine his vast and rich opus may, of course, find passages in support of every one of these contentions. Like any great figure, Augustine displayed many sides in his thought, and different interpreters have emphasized different elements.

In much religious literature on the environment, Augustine is persona non grata; he doesn't seem a good example of the embodied engagement with nature many today seek. But a reading of Augustine for what he says about creatures and creation would turn up some surprises. In the last three books of the *Confessions,* where he deals with the creation stories in Genesis, it is quite clear that Augustine, like the whole orthodox Christian tradition, regards nature as a manifestation to us of God's Wisdom. In fact, in that tradition, he links nature and revelation as twin sources for our knowledge of ourselves and God. In one passage (xii, 15), for example, he says to God that the angels do not need to look upon the firmament or read to attain knowledge of his Word, but the clear implication is that we humans do. He argues for a charitable openness and tolerance for all the legitimate truths that may be drawn from the story of creation in Genesis but proposes some interpretations himself of how the Trinity, man, and nature are interrelated.

Augustine's, then, is no mere abstract biblical cosmology; he links the cosmos and man. The Trinity is an image of our existence, he argues: the Father is Being, the Son is Wisdom or Intellect, and the Holy Spirit, Will. And being, knowing, and willing, united somehow in each of us make us an *imago Dei* (xiii, 11). Our ability to know things, to make objects, and to perform acts is a reflection of the divine life. Augustine even mentions Wisdom, not in the current feminist mode, but in a central position all the same. For Augustine "Wisdom is Thy Son in whom thou hast created Heaven and Earth" (xii, 5).[12] Far from minimizing creation, his curiosity about the world leads him to probe one of the trickiest questions of all: the nature of time. Modern philosophers and scientists have wrestled with that notion as well, but there is no reason to regard Augustine's speculations about time as inferior to Bergson's or Heidegger's, Einstein's or Hawking's.

In the *Confessions,* Augustine says that the omnipotent God envisaged in the Bible had to create all things from nothing, because if he had made them, as we make things, from already existing material, he would not be the all powerful. And Augustine adds an astonishing insight: we shouldn't think of God as making things appear in heaven or on the earth. "How, O God, didst thou make heaven and earth? Truly, neither in heaven nor in the earth didst Thou make heaven and earth; nor in the whole world didst Thou make the whole world; because there was no place wherein it could be made before it was made" (xi, 5). As relativity theory and modern cosmologies maintain, the universe we know was not simply an explosion into a pre-existing void but was the creation of space as well as objects. The discovery of the big bang was a shock to the old classical physics, but Augustine could quite easily have understood the general import of modern cosmology because, like a Stephen Hawking, he understands that creation also creates time. Thus for Augustine, time and space are not merely the neutral matrix for all events, as was mistakenly thought during the rise of modern science. They are the basis for an ongoing, unfinished story that involves the human race and the whole universe. God must also sustain everything at every point; otherwise it would fall back, as certain quantum phenomena do, into nothingness. Whatever happens in the universe lies along the twin axes of creation and continuing providence. Or as Abraham Joshua Heschel once put it, the whole of Hebrew Scripture is an "architecture of time."[13]

Augustine was the great expositor of the biblical emphasis on history: God acted through time from Creation, through the Fall of man, the election and wanderings of the Israelites, down to the culmination in Christ's incarnation, death, and resurrection. The notion of linear progression in time — personal, communal, and cosmic — though not entirely new with Augustine, was to give Christianity a this-worldly dynamic that would make it both different from paganism and, in a way, temporally fruitful.[14] Augustine speaks of *rationes seminales,* or seminal energies, that God has hidden in nature almost like the seeds of a tree. These make possible novelty, discoveries, and invention. Certainly the West's confidence in exploring hidden reasons and seeking to use them in the hope of furthering the divine plan derives in part from what Augustine discerned in the Bible.

The Straight Way

Some critics see all this as the heart of the West's environmental problem. For them, nature and pagan religions seem better adjusted to each other because they both, it is thought, move in repetitive cycles. In our time, when technologies have blunted our sense of nature and our respect for time and the seasons, Augustine may appear a forerunner of the kind of narrow linear thinking that really only emerged in the eighteenth century with the Enlightenment. But the strongest vein in his work combines appreciation for God's overflowing goodness in the immense proliferation of great and small creatures in nature with gratitude for his liberation of the human race from false idolatries regarding nature. Among those idolatries, though Augustine praises the round of the seasons, are pagan cyclical theories, a terrible prison for the soul:

> Let us therefore keep to the straight path, which is Christ, and, with Him as our Guide and Savior, let us turn away in heart and mind from the unreal and futile cycles of the godless. Porphyry, Platonist though he was, abjured the opinion of his school, that in these cycles souls are ceaselessly passing away and returning, either being struck with the extravagance of the idea, or sobered by his knowledge of Christianity. (*City of God*, xii, 20)[15]

This is the locus classicus for the Western rejection of cyclical time and the embrace of history. Though the "straight way" of Christ refers primarily here to a spiritual liberation, Augustine goes on to remark that God creates novelties even in nature (a view that modern cosmology seems to have confirmed). Augustine's straight way explodes the notion that we must endure endless cyclical repetition whether in nature or the spiritual life. Today, we may see the denial of cycles as inviting ecological disaster, but it is no small evidence that millions of souls in antiquity, while continuing to honor and reflect the cycles of nature in the Christian liturgical year and in cosmological ritual, nevertheless found the linear time of Christ a liberation from paganism.

The Human Factor

If we think that human beings are created in God's image and likeness and that God works though history, we will not be able to acquiesce in a steady-state view of nature that both revelation and scientific research have denied. Augustine elsewhere speaks of God as marvelously creating, ordering, guiding, and arranging all things "like the great melody of some ineffable composer."[16] New things are always being added by God to that cosmic symphony. And it is the very essence of our own emergence from the cosmological process that leads us to continue cosmic development in various ways. For example, we turn trees into wood, wood into violins and other instruments, instruments into orchestras that play Beethoven. In a similar way, the benefits and developments that human creativity has been able to draw from nature cannot be seen simply as a denial of God's order or a threat to nature (although, unwisely pursued, they may become so), for in the longer run they may help us to realize in nature goods for ourselves, for animals and plants, and for the universe, which it seems God intended not to exist without us. Abraham Lincoln, that naturally great and pious soul, once remarked: "All nature — the whole world, material, moral, and intellectual — is a mine; and in Adam's day, it was a wholly unexplored mine."[17] We might not put it quite that way today, but Lincoln is onto a sound Augustinian point.

When all this has been said, however, Augustine parts ways with his modern critics over some quite definite issues. He is not, as they are, inclined to think that the splendors of creation almost automatically override the evils of the Fall. He is wary about the urge to know and possess "natural pleasures," citing the first chapter of Paul's Letter to the Romans as an indication that some people exchange natural ways for unnatural ones without acknowledging as much. He is also concerned to preserve the right — which is to say divinely willed — order in our appreciation of nature. And in the last analysis, he thinks it necessary to maintain a hierarchy of goods, including the good of human nature. His famous saying "My love is my weight" appears in a context where he shows the wisdom of the Creator in the way each thing seeks its proper place. Our place in the cosmos involves our love of God and neighbor; we do not merely occupy a physical position like inanimate objects, nor follow strictly determined instinctual behavior as do plants and animals. Furthermore, the world, he says, and all the things in it are very good (xiii, 30), but they are not as good as God himself (xiii, 31), the ultimate end of all existence.

In the opening pages of the *City of God,* Augustine acknowledges that many things that occur naturally — earthquakes, the death of

children, crippling diseases—provide ammunition for the pagan critics of a Christianity that believes in the goodness of the creator. He
does not claim to know why God allows such evils but notes that
there are several plausible psychological and spiritual explanations. If
believers always received good and unbelievers evil, people might begin converting out of self-interest rather than true love. Furthermore, natural ills convert us from our own self-will and intellectual
pride. We have not made ourselves and we cannot be the masters of
our own bodies, let alone nature, for long. Many otherwise upright
people need to be weaned away by God from this false sense of their
own righteousness and power, or from a superstitious belief that nature is God.

Some of these arguments remain persuasive, others less so, but
Augustine is actually being more realistic about nature here than are
many of his critics. He appreciates the wonders of creation, but he
never allows himself to fall over into the facile romanticism that expects us to achieve a kind of return to the Garden by getting back in
tune with nature.

But Augustine alone of Christian thinkers up until his time maintained that the cosmos had not fallen along with the human Fall.[18]
When Saint Paul says in Romans 8 that creation had been subjected
to futility and was groaning in travail waiting to be restored, Augustine argues that this refers to the human creature, not nature in general. He may have been concerned to refute a position quite common
in the classical world in the Gnostic and Manichean religions. Gnostics thought, seeing evil and death, that the creator must be an evil
Demiurge, and the Good God was to be found only beyond nature.
Manicheans found Good and Evil principles at war in this world.
Both, therefore, allowed absolute evil a foothold in the very structure
of existence. Augustine was a Manichean in his youth but later became very sensitive to any hint that some pretemporal evil was mixed
into the world. When he became a Christian, he had to deny that
completely. Certainly the passage in Paul is far less sanguine about
the goodness of present nature than much that is found in Augustine.

Still, with the whole Christian tradition, he does not expect reversion to an idyllic state; that can come only at the end of time. Augustine would never oppose alleviating the human sufferings that he regards as the effects of the Fall. But he would be skeptical of any

promise—whether through retreat to Eden or through technological wonders — that gave the impression that we can overcome sin and death by any other way than repentance and submission. He chides the Stoics, who thought nature divine, for also thinking suicide the remedy for unbearable ills: what kind of divine life is it, he asks, if sometimes we ought to kill ourselves?[19]

Because of the limits Augustine sees in this life, some modern theologians have argued that he had an otherworldly hatred of the body — a deeply mistaken reading of the mature Augustine. Along with his acceptance of human dominion over creation, they see in him the deep roots of an exaggerated spirituality that leads to the exploitation of nature. But in Augustine dominion primarily means that the human being "judges all things."[20] And anyone who has read Augustine with care will recognize that an aggressive pursuit of wealth or material goods is simply alien to all phases of his thought. Still less is sheer domination a good for the writer who made *libido dominandi* (lust for domination) one of the main targets of his criticism of hidden forms of pride and injustice in ancient Rome. Augustine is more properly to be considered one of the Christian sources of a simultaneously restrained and creative, spirit-filled respect and use, not abuse, of God's creation.

The Estimable Cardinal

Augustine exposes some of the deep and fertile insights to be found in the biblical vision of the world; those insights took various directions in the course of Western history. For an orthodox Christian evaluation of that history and its relevance to ecological issues, we need to turn to a modern commentator, Joseph Cardinal Ratzinger, a distinguished theologian who since 1981 has been Prefect of the Congregation for the Doctrine of the Faith in the Vatican. In past centuries, his congregation ran the Roman Inquisition, and that fact, along with Ratzinger's orthodox Christian views, has made him the target of much criticism. But before assuming his post as the chief expositor of Christian theology at the Holy See, Ratzinger gave a series of four homilies at the Liebfrauenkirche in Munich because, as he later explained, "the human threat to all living things, which is being spoken of everywhere these days, has given a new urgency to the theme of Creation." Yet the doctrine of Creation appeared almost nowhere in contemporary catechesis, preaching, or even theology.[21] Ratzinger lectured before the explosion of religious environmental literature, but it is doubtful whether that development is what he hoped to see.

In Ratzinger's view, the typical modern theologian appears almost embarrassed by the idea of the creation. Citing several prominent figures, he suggests that they feel far more comfortable with scientific mechanism as the explanation for the universe, while they tend to regard the concept of creation as existential or mythological. Of course, these thinkers still believe human beings find their deepest meaning and source in God, but creation—on the rare occasions when the term appears—necessarily has a changed meaning in their work, and certainly does not much help with our profound confusions about humanity's place in this world and the obligations we may have toward it.

It may not be entirely by chance that Ratzinger dated the preface to his book *In the Beginning . . .* on the feast of Saint Augustine. Ratzinger seems to suggest by this subtle juxtaposition that, for all the criticism directed at Augustine, he belongs at the center of the discussion. (Augustine wrote fully or in part at least five commentaries on Genesis in various works). By contrast, contemporary theologians have almost forbidden themselves, out of some embarrassed need to be absolutely modern, to resort to the very portions of the biblical and theological tradition that may promise the most in confronting environmental questions.

The opening of Genesis, Ratzinger says, impresses almost everyone with its poetic evocation of the seven great "days" of Creation. But can we say of this account that it is "true" anymore? The old Newtonian view of the universe made it incredible to many people over centuries. Complicated contemporary notions of time and space, Einsteinian time-space, quantum theory, the big bang, and chaos theory have all enormously expanded our picture of the universe. Compared with either system, the biblical account seems to belong to the infancy of the human race, and some theologians call for abandoning the Creation myth and accepting purely scientific hypotheses for the world. But, asks Ratzinger, is there another way to read these stirring chapters of the Bible that would avoid relegating them to the category of mere nostalgia?

He answers that there is, but that we must be careful to understand that this is not merely a later rationalization undertaken under challenge from modern science. In the first place, as virtually all the traditional commentators since the time of ancient Israel have acknowledged, there is a difference between the form and content of the Creation story in Genesis. For some modern readers, this is a virtual confession that the Bible has no solid truth value, merely a symbolic one that can be changed for whatever purpose we wish. In the English-speaking world, skeptics like Bertrand Russell have asked why the all-powerful God, who is apparently so concerned for truth, did not simply tell the early Israelites the correct, "scientific" account. And popularizers like Will Durant, following Spinoza, have claimed that the Bible is full of Oriental exaggeration and symbolism directed toward emotion, not reason.[22]

Ratzinger counters that if that is all there is to the Creation story, then the same technique can be made to redefine everything in the Bible including miracles, the incarnation, and Jesus' cross and resurrection. Faith itself hangs in the balance, and honesty demands that we confront the whole question. Many theologians who do so wind up with a "sickly Christianity . . . that is no longer true to itself and that consequently cannot radiate encouragement and enthusiasm."[23] Ratzinger's answer to this predicament is that the account of Creation in Genesis was never closed in on itself from the beginning and should not be mistakenly approached as if it were "a novel or a textbook" — intended either as pure poetry or as literal truth.

The Biblical Difference

Instead, the Judeo-Christian notion of God as a person intimately involved in the history of his people provides a far different starting point for thinking about nature, and one with impressive, worldwide consequences. Just as we may glimpse in the history of Israel how God gradually has been leading us forward to greater understanding, so we must understand the text of Genesis as providing the beginning of an intellectual and moral journey, first of the Chosen People, and later of the whole human race. Similarities between the biblical account of Creation and those in every other great culture, even those with no contact with one another, suggest that God never lost touch entirely with any part of the human race. But the Creation account in the Bible is different from the others. Historically, it has proven its staying power as circumstances at given moments demanded.

One particular moment uncovers differences between the Bible and other sacred cosmologies. During the Babylonian Captivity, the Chosen People were forced to think about Creation again. The powerful Babylonian empire — and its gods — appeared to have triumphed over the God of Israel. Yahweh, the God of the tribe, had to be seen again as powerful everywhere, even in exile. And reflection on that fact introduced an element central to several future religious developments. The Babylonians believed that the universe was created when the god Marduk slew a dragon called Tiamat and used parts of the body to create the universe. The two opposing parts of Tiamat, struggling together, gave rise to the mixture of good and evil we experience in matter. In particular, human beings emerged from the dragon's blood, which meant that something sinister and dragonlike lay in all things and in the very heart of human beings — and not as an addition like sin but in our basic constitution.[24]

Though Ratzinger does not mention it, in the distant Americas, Mayan and Aztec cultures developed a similar notion of the "blood of the gods," and the need for human blood sacrifices formed the deepest strata of Meso-American cosmology. As the Mexican novelist Carlos Fuentes, no apologist for Christianity, has put it, we are all so familiar with the biblical Creation account that it has made us blind to what it replaced and corrected in other cultures. Particularly with regard to the conversions in Meso-America, Fuentes has observed:

> One can only imagine the astonishment of the hundreds and thousands of Indians who asked for baptism as they came to realize that they were being asked to adore a god who sacrificed himself for men instead of asking men to sacrifice themselves to gods, as the Aztec religion demanded.[25]

The biblical account that in many places replaced others emphasized the complete sovereignty of a benevolent God over creation.

In both the ancient world and around the globe, some novel and important consequences ensued. Sun, moon, stars, and the earth itself were no longer gods or demons, merely fellow creatures. Many people in many cultures must have at first thought this a great blasphemy. We hear an echo of that reaction in contemporary environmentalist laments about the crushing of nature worship and polytheistic nature cults by biblical monotheism. But for large parts of the world, that monotheism was an "enlightenment," and a true one. Furthermore, it came directly from God's walk through history with his people and therefore reappears at crucial historical junctures, like our own. By the time Saint John wrote at the beginning of his gospel, "In the Beginning was the Word, and the Word was with God, and the Word was God. . . . All things were made through him, and without him was not anything made that was made," Christ himself had been identified with a view of creation that over time would overcome all the gods and goddesses, monsters and mythological beasts of Egypt, the Middle East, Greece, and far beyond.

The Biblical Mode of Rationality

Ratzinger argues that the ancient and medieval church knew that this was the right "rationality" to be found in the Bible. Only modern forms of historical thought have tried to read every text through a bare literalism. For him, the harsh exploitation and domination of the world by human reason derive from a methodology that has arisen only over the last few centuries and is alien to the biblical worldview:

> The reasonableness of the universe provides us with access to God's Reason, and the Bible is and continues to be the true "enlightenment," which has given the world over to human reason and not to exploitation by human beings, because it opened reason to God's truth and love. Therefore we must not in our own day conceal our faith in creation. We may not conceal it, for only if it is true that the universe comes from freedom, love, and reason, and that these are real underlying powers, can we trust one another, go forward into the future, and live as human beings. . . . For this means that freedom and love are not ineffectual ideas but rather that they are sustaining forces of reality.[26]

Nonbiblical reason, despite its practical success, deformed human thought in the past few centuries. In the nineteenth century, for example, most educated people came to believe that the world was the result of random accidents, not the passage from a beginning to an end, as in the biblical story. But as science has advanced, we all now recognize that the world had some sort of origin, though the end point of creation still remains hidden and disputed. Ratzinger asserts that "the more we know of the universe the more profoundly we are stuck by a Reason whose ways we can only contemplate with astonishment."[27] This is not mere pious sentiment but a plausible and accurate account of the wonder of advancing scientific explanation itself.

Of course, biblical thought conceives of creation in a different way than do modern physics and biology. For human beings, morals and piety are part of the very fabric of creation, a fact universally recognized, if in different forms, by all civilizations. When piety towards God's creation is kept pure, it reflects a dependency in love between creature and creator. Where it takes less pure forms, it is thought that the gods need something from human beings and that their favor can be bought through all sorts of magical rites and practices. The pure impulse results in what Ratzinger calls the "Sabbath Structure of Creation." The Sabbath is the seventh day, when God himself rested and saw that creation was good. In human terms, the Sabbath every seven days is not only a time for rest from labor and a turning of the mind toward God but also a return to the beginning, a sloughing off of the labors of the everyday and their disfigurements, and a beginning anew from the source of everything. The Sabbath is thus not so much a rest as renewal and return to the point of primordial joy in existence.

In an ingenious reading of several complex biblical passages, Ratzinger suggests that the vehemence of the prophets against Israel had to do not only with violations of the Ten Commandments, as serious as these were and are. Instead, he says, citing a passage from the Second Book of Chronicles, when the Israelites forgot the Sabbath, the land still celebrated it: "Until this land has enjoyed its Sabbath rest, until seventy years have gone by, it will keep Sabbath throughout the days of its desolation." The ancient Israelites thus fell not only into sin but also into a rejection of their original relationship with God, a rejection that took a particular form: "the people had rejected God's rest, its leisure, its worship, its peace, and its freedom, and so they fell into the slavery of activity."[28]

Some modern critics have argued that hyperactivity was, and continues to be, the logical consequence of the Bible's own demythologizing of nature and the rationalization that came from the biblical revelation, including the command to subdue the earth and all the ecological consequences that seem to flow from it. A world formerly full of spirits became merely utilitarian material. Yet Ratzinger counters that the Sabbath vision never allows us to be merely closed in on ourselves, but opens us out to past and future and to an expansive vision of what creatureliness entails.

It is much easier to trace, he argues, the modern mentality in the great figures who departed from the central Christian tenets. Galileo believed nature could be tortured to extract its secrets.[29] By the time Marx appeared, the whole question of origins was simply mocked: we have no way of knowing nature's origin, Marx thought; therefore we are free to do what we wish with nature, including human nature and society, for our own benefit. Hence the notorious ruthlessness of those seeking to establish the communist society and, we might add from what we have learned about communist ecological destruction in the years since Ratzinger wrote, the stunning environmental devastation of communist nations. The nonbeliever Ernst Bloch argued similarly that progress is the only truth and that human beings would, out of their own values, forge a world worth living in. In the future, according to Bloch (echoing strains of Henry Adams's Dynamo), laboratories and power plants will become our cathedrals and basilicas.[30] These rhapsodic accounts of human activities seem outdated, but the most progressive minds were proposing such absurdities just a few decades ago, and some still do today.

A Humble and Exalted Kingdom

More commonly, we now see this kind of human being himself as the wrecker of nature. When people sympathize with whales or snail darters instead of human beings today, it may be because of an intuition that if this is what human beings are, the only solution to the ecological problem seems to be for human beings to put themselves out of the way. And that is precisely what several different schools of environmental thought have proposed. That these are all versions of suicide, says Ratzinger, reveals the love of death and the self-hating impulse that inevitably flow from the attempt by modern utilitarian forces to set up a kingdom separate from God's, a kingdom whose traditional name is Hell.

As a counterweight to this demonic view, Ratzinger returns to the biblical account of the creation of man. That account is quite humbling—deliberately so. All men and women of every race and nation are formed of the "dust of the Earth." No social distinction, no ideology of "blood and soil" like the Nazi idea, can outweigh this basic biblical notion of fraternity in our material origins. The earthy part of the creation account comes first and underlies the traditional notion that grace builds on and presupposes nature. The second element in human creation, God's breath breathed into the dust to give life to the man Adam, suggests two things. First, humanity is one, but each person is individually created and called by name. All attempts to crush out that individuality in the name of the human mass actually constitute a violation of God's ownership of each of us: "When the human person is no longer seen as standing under God's protection and bearing God's breath, then the human being begins to be viewed in utilitarian fashion."[31]

In the context of the long-running debate on creationism versus evolution, the breathing of God's spirit into the dust explains *what* man is, not how he came to be. Both creation and evolution may be seen as playing a role in the final result. Taking the French biologist Jacques Monod as the best and most powerful modern proponent of a strict evolutionary model, Ratzinger points out that older nineteenth-century views of the necessary emergence of physical states from previous states, as in Laplace or Hegel, are not regarded, even by scientists, as tenable in this century. In addition to the complications introduced by quantum theory and other modern discoveries, we also know how sheerly improbable life, particularly human life, is if the universe emerged purely by chance. Monod himself contends that our improbable emergence from nature is an unbelievable windfall: we are like people who have bought a lottery ticket and unexpectedly won a billion dollars.

In Christian language, Monod has here identified the "radical contingency" of the human being. Living organisms may be infinitely more complex than any machine and contain within them potentialities that no machine possesses, such as self-movement and the power of reproduction. Still they remain nonessential in any physical sense, the mere product of happenstance. To this, Ratzinger responds that this is neither the way in which we perceive ourselves nor the direction in which developing science is pointing. Late-twentieth-century physics is more and more likely to emphasize reason built into the universe from the first instants of creation and pointing towards us. Oddly, while physics has been emphasizing the interlocking rationality of the cosmos, biology continues to believe randomness drives evolution.

The Return to Truth

In a concluding chapter, Ratzinger hints that the denial of our place in nature has a great deal to do with sin, or, as Christopher Derrick has argued, not only do we need to correct our view of nature as a mere machine, but also "we need to repent of a heresy."[32] Through sin, men have always denied God and his sovereignty, but that denial has led in our time to a misconstrual both of the environmental problem and of the potential solutions to it. Recognizing our error is difficult. Ratzinger argues that in our current circumstances, we hunger for a healing relationship with creation but can arrive at it only if we both become true in our very ways of living and stop suppressing or destroying the truth.

As an approach to rediscovering truth, we need to rethink our history since the Garden of Eden. The earth is a temporary home, a gift, but it remains the property of God. Any attempt to use it for some order outside God's own order will result in disaster. In a sense, the temptation of the serpent at the Fall was precisely a call to an order separate from God's, or in the words of biblical scholar G. Von Rad, "the possibility of an extension of human existence beyond the limits set for it by God."[33] The serpent's temptation is not a direct denial of God, merely a questioning of God's commands that ends in his suggestion that Adam and Eve not be bound by God's limitations, particularly on what constitutes good and evil.

The rupture caused by Adam and Eve's decision has had many historical consequences, but Ratzinger emphasizes two contemporary errors: the aesthetic and the technical. He finds in modernist aesthetics the belief that what constitutes the beautiful is what is well executed, whatever other negative dimensions it may have. Beneath this assumption, he says, lies an anthropology that man is defined by what is within his capacity to do: "if there is an area where human beings can ignore their limitations: when they create art, then they may do what they can do; they have no limitations. And that means in turn that the measure of human beings is what they can do and not what they are, not what is good or bad. What they can do they may do."[34]

The second modern deformation, the technical, says Ratzinger is a more evident threat. But he points out that the Greek word *techne* bears a meaning very close to the modern English word *art*. And there is some deep connection between the belief that man may do what he wants in art and the belief that he may do what he wants with science. J. Robert Oppenheimer expressed some moving doubts — after the fact — about having helped to create the atomic bomb. But at the time, he was intrigued, as a scientist, by the technical problem. Rudolf Höss, commandant at Auschwitz, could speak of the camp as a "remarkable technical achievement." In these well-known examples, mere technical means, absent reference to the standard of good and evil and God's creation, do not free us. On the contrary, what appears a liberating technique becomes a means by which men "are destroying themselves and the world."[35]

The notion of original sin has sometimes been used to suggest that, almost by an automatic process, the original fault has been spread to the whole race. Ratzinger points out that there are actually three dimensions to this process: we lose our connection to God, to one another, and to creation. And all these dimensions show themselves in human action throughout history. The only remedy is to recognize our creatureliness as our true nature and, in our personal lives and public actions, to repair the damage done. With regard to creation, he says, "Since the relationship with creation has been damaged, only the Creator himself can be our savior." And in the New Testament account of Jesus, we see the Adam story in reverse: a man willing to endure the disorders and temptations of the broken relationship with God in order to show the right relationship to God, creation, and other human beings—the tree of life lost at the Fall and increasingly eclipsed by a culture of death.

In environmental terms, the science and technology that the nineteenth century boasted would increase human happiness and inevitably replace religion have themselves become a possible threat to human flourishing and even existence. One answer to this problem is to propose purely technical remedies — a gnostic belief that pure knowledge will save us. Certainly, better and more sensitive use of technology is called for. But Ratzinger takes a more radical stance: " 'What can we do?' will be a false and pernicious question while we refrain from asking, 'Who are we?' " The question of our being and the question of our hopes are inseparable."[36] And he continues:

> To go straight to the point: the foundations of modernity are the reason for the disappearance of "creation" from the horizons of historically influential thought. Thus our subject leads us to the very center of the drama of modernity and to the core of the present crisis — the crisis of modern consciousness.

The Three Concealments

Ratzinger brilliantly identifies three forms of the concealment of creation in modern life and thought. The first is the modern scientific view of nature. As useful as it is for some purposes, once we turn to "human nature" or "human rights," this view instantly reveals its uselessness. No amount of physical or biological knowledge can establish or help sustain these moral and political notions. A kind of schizophrenia results in which we talk about things we believe to be right, but in the void of a universe that seems to give them no place to stand: "If creation cannot be recognized as the metaphysical middle term between nature and artificiality, then the plunge into nothingness is unavoidable."[37] In many ways, all the current forms of postmodernism recognize that the abyss is the result of our present-day conceptions of human life vis-à-vis nature.

A second concealment stems from the reaction against science and technology. Rousseau may have been the first to object to technological reason and to oppose to it what he thought to be pure nature, but since his time thought has undergone a further development. Nature, as some see it, is a beautiful system whose balance man continually disrupts. Both the human mind and human freedom are diseases, breaks with nature scientifically viewed. If there is any healing, it is in a return to anthropological simplicities, as in Claude Lévi-

Strauss, or to strict animal behavior, as in B. F. Skinner. In either case, the result for human beings is a nullifying of the specifically human, or, as has been said in a variety of contexts, nihilism.

Finally, there is a theological concealment of creation in what Ratzinger calls a "monism of grace." Contrary to the older biblical view of the human as part earth, part spirit, one reaction to our predicament is the *odium generis humani,* hatred of the human race. Ratzinger identifies this with the various forms of gnosticism that despise, or seek to be free of, the body and creation. In a sense all the modern forms of concealment are gnostic because they put false knowledge in the place of humble acceptance of creation:

> Love appears too insecure a foundation for life and world. It means one has to depend on something unpredictable and unenforceable, something we cannot certainly make for ourselves, but can only wait and receive. . . . The Gnostic option aims at knowledge, and at power through knowledge, the only reliable redemption of humankind. Gnosticism will not entrust itself to a world already created, but only to a world still to be created. There is no need for trust, only skill.[38]

Ratzinger would not deny that many goods lie in the future because of human innovation. Our creativity, however, must reflect our creatureliness; otherwise our creation is destruction. Human action must stand under redemption:

> Only if the being of creation is good, only if trust in being is fundamentally justified, are humans at all redeemable. Only if the Redeemer is also Creator can he really be Redeemer. That is why the question of what we do is decided by the ground of what we are. We can win the future only if we do not lose creation.[39]

Four Unresolved Questions

Even in this brief treatment, these ideas reveal deep meditation by a powerful mind. Yet there are several topics Ratzinger was not able to include in these lectures. Some we have already seen in Augustine; others emerge more starkly from practical choices we face. Basically, these revolve around four issues.

First, creation in itself may be good, but man experiences its threats and challenges as, at least in part, evil. Whether we think of diseases or natural disasters like earthquakes, floods, or hurricanes, there is something in nature other than simple goodness to be taken into account. After the 1755 earthquake in Lisbon, Voltaire—following in the footsteps of the pagans who mocked Augustine's belief in the goodness of the world—pronounced in *Candide* that the tens of thousands dead once and for all disposed of the notion of an all-provident God. A believer who seriously thinks about such events has to admit that the goodness of creation is both a basic datum of revelation and a difficult question. The creation may be good, but its goodness is partly unintelligible to us.

A second issue flows from the first. We necessarily both contemplate nature as God's gift and grapple with it as a challenge. Our capacity to discover potential in matter and repel threats to ourselves is not mere Promethean rebellion against God's order. Our very action may be, in the right circumstances, cooperation with the original creative impulse from God, restoring the part of creation damaged by the Fall and developing the part of creation that unfolds through time at God's behest. As we have seen in Augustine's rejection of Manicheism, something of a progressive dynamic seems to be built into the structure of the universe, and part of recognizing human creatureliness is to allow for the proper use of the powers God has given us.

A third issue thus emerges. Man is both a contemplative and an actor, and his creation in God's image is most evident when both of these dimensions are in the right relationship — sometimes of greater activity, sometimes of less. The Platonizing side of the Christian tradition rightly emphasized that our ultimate good lies beyond the world, but this emphasis sometimes came at the expense of the undeniable goods of practical action. Ratzinger rightly returns us to the fuller tradition that grace builds on nature, but he leaves open the question of how activity in this world and contemplation are related in light of the grace-nature relationship. In particular, the notion of the limits God has placed to human action within his created order needs to be spelled out more fully. We know that, morally and practically, there are such limits. But particularly in practice, we should not be too quick to assume that all we are permitted to do is to maintain some pre-established order — something that neither the Bible nor the scientific picture of an unfolding cosmos permits us to believe.

Finally, Ratzinger puts the human creature in its proper place: we should not flee full humanness or the properly human world for a kind of animal existence in an environment; neither should we try to establish a sufficiency separate from God through a gnostic technological transcendence of the world. The developing powers of human science and technology call for a new creation theology. Environmentalism and ecology, at least as usually practiced, are too shallow and weak for this task. Environment, as we have seen, is a notion that applies more strictly to animals, which respond to stimuli in an environment rather than possessing a world. Ecology might be a better term, if we took it in its etymological sense as the rationality of the household, meaning the planet and the entire creation. But the word ecology has become associated with pragmatic approaches or theologies that all but divinize nature. Therefore, we need to work out a creation theology and a practice thereof that will gradually heal the split between our knowledge and power over the physical realm, on the one hand, and the fullness of our moral and spiritual experience, on the other.

If we lose that perspective and the crucial biblical heritage that opened up our abilities to study creation in the first place, we will lose the ends that the technologies are the means to serve. As the philosopher Leszek Kolakowski has characterized the modern predicament:

> It appears as if we suddenly woke up to perceive things which the humble, and not necessarily highly educated, priests have been seeing — and warning us about — for three centuries and which they have repeatedly denounced in their Sunday sermons. They kept telling their flocks that a world that has forgotten God has forgotten the very distinction between good and evil and has made human life meaningless, sunk into nihilism. Now, proudly stuffed with our sociological, historical, anthropological, and philosophical knowledge, we discover the same simple wisdom, which we try to express in a more sophisticated idiom.[40]

If there were no intellectually respectable reaction to modern science but to submit and accept such meaninglessness, that would be that. But science itself has entered a complicated new phase that will form the subject of the next chapter.

2

A Dull Child's
Guide to the Cosmos

The growing convergence between religion and science is so remarkable that it has become a common topic in popular magazines. In one week during the summer of 1998, for example, both *Newsweek* and *U.S. News & World Report* carried cover stories about God and modern cosmology.[1] Large questions about the nature of reality based on the work of Einstein, Hawking, sophisticated cosmology, and cutting-edge particle physics make up the background of these stories. Yet for all the coverage, popular consciousness of these discoveries remains largely uninformed. And even more remarkably, there has been little attention to what some of these discoveries may tell us about life on the planet or about the environment. Scientists have begun to connect the two, the Virgin and the Dynamo, and it will be useful if we look here at some of their findings, beginning as all such surveys must with the crucial figure behind the modern scientific worldview.

In the first decade of the twentieth century, something unprecedented in the whole history of the universe occurred. Albert Einstein, then an obscure worker in a patent office in Bern, Switzerland, made several remarkable discoveries in his "spare time." The most important was what later came to be called his special theory of relativity. Trying to reconcile several apparently contradictory data about matter and electromagnetic effects, Einstein hypothesized that time and space were part of a single continuum and were not absolutes, but could be altered when objects were moving at speeds close to the velocity of light. The ramifications of this theory, which has been confirmed by various experiments since it was first published, are still being sorted out. But the human significance of this event is even greater than the scientific achievement: for the first time since the origin of the universe (so far as we know) one of the life forms to which the universe had given rise was able to look, from a vantage point on an ordinary planet circling an average star, across the vast cosmic times and distances and comprehend how they all relate to one another. In other words, a tiny portion of the cosmos was capable of taking in the whole, at least from one perspective, in a way that was not merely symbolic or mythological but that encapsulated, with great precision, everything.

Einstein's work has not yet been much absorbed into popular consciousness. In fact, his theory of how all things are related has been culturally twisted to mean that everything is relative. But Einstein, then and in his subsequent work, was really explaining how different frames of reference stand towards one another. And his insight has relevance not just for physics but for the kinds of reflections we are engaged in here as well. It may seem that cutting-edge physical science takes us far afield from religion and environmentalism, but, as we saw about theology in the previous chapter, if we really want to know what we are about at our present moment, we need some very large intellectual perspectives. About the same time that Einstein was making his first discoveries, Henry Adams thought that the Dynamo of science and technology had forever vanquished the world represented by the Virgin of religion and art. People like Adams could deeply lament the loss of something they felt humanly valuable beyond all calculation (and might even secretly pray, as Adams did, to what they thought was a product of the human imagination). But they could not deny the evidence produced by human reasoning about the mechanical workings of the universe. In the confrontation between the Virgin and the Dynamo, only sentiment seemed to side with the Virgin. Truth, reality, and the future itself seemed to favor the Dynamo. But with Einstein and other great figures of twentieth-century science, the Dynamo began to look a lot more like the Virgin.

Settling an Old Quarrel?

Most people today have the vague impression that science and religion remain opposed to each other, with science delivering all sorts of goods in ways that have weakened religion. Even though scientists, particularly astronomers and physicists, have in recent years been increasingly eager to talk about the possible religious implications of their work, the negative images from the Galileo case and the ongoing Darwinian controversies continue to color our thinking. Some scientists continue to attack religion as a dangerous superstition that needs outright social regulation;[2] others — slightly under half in recent surveys — openly admit to being believers.[3] Religious people have not particularly helped the situation. Some of them stubbornly regard what science has been doing as something like the product of the devil. Others try to seem absolutely up-to-date by conceding everything to scientific views that instead need critical religious reflection, especially in light of their impact on human morality and meaning. Reason seems opposed to faith; and since science, for us, largely represents reason, religion often appears, at best, well-meant irrationality.

And yet if science tells us some important truths about the world, we cannot advance any further in examining the relevance of religion to environmental issues unless we digest at least a few spoonfuls of the best thinking about the current relationship of religion and science. As Einstein himself remarked, "science without religion is lame, and religion without science is blind."[4] This is an enormous issue that can be treated only briefly here, and the reader who acquired an allergy to science in high school might want to turn immediately to the next chapter. Much of modern science depends upon mathematics of a very high order, which few of us are skilled in understanding. But without entering into those mathematical and scientific complexities, we can still get an accurate, if oversimplified, idea of some of the crucial questions concerning the origins of the universe, its unfolding through time, the nature of life (including human life), and what all these things, as currently understood in modern scientific thought, may tell us about our relationship to the environment.

We desperately need our religious commitments to be illuminated by accurate knowledge about the world, which has begun to look quite different than we once thought. Prior to modern times, many theologians viewed matters quite differently than we do. Thomas Aquinas, certainly one of the greatest of all Christian theologians, saw science and theology as two parts of one human process of knowing. He asserts boldly that "errors made about creatures sometimes will lead one astray from the true faith."[5] And, citing Psalms 28, he even rebukes those who think that it does not matter what we think about creatures so long as we believe the right things about God: "Because they have not understood the works of the Lord and the work of God's hands, you will destroy them, and not build them up" Indeed, Aquinas maintains that "to detract from the perfection of creatures is to detract from divine power."[6]

Not all Christian theologians would agree with these points by any means, but Aquinas is clearly enunciating a central strand in the biblical tradition that brings science, properly understood, within the act of faith, rather than viewing it as a necessary challenger or something independent of faith. The Christian life must be a unity; there cannot be one truth in science, another in theology. Aquinas regards science as essential for some purposes and a possible delusion for others: "Sciences ensure right judgement about creatures. Their drawbacks are that they are occasions of our turning away from God."[7] For those of us who do not wish to do without either half of our human powers of knowing, this invites careful reflection, especially about the significance of current physical science.

That reflection is once again fruitful because there has been a strange convergence recently among physics, astrophysics, and what seemed at one time obsolete elements in the Christian story of Creation. The idea of a Creator is a complicated notion, since Creation is usually taken to mean that an all-powerful God made something from nothing *(ex nihilo)*. Even without scientific qualification, the sheer difficulty of the problems contained within the idea of Creation is enormous. How can something come from nothing? If God is all-powerful, does his domination consist solely in the initial shaping of the universe, or may he intervene and change the laws he seems to have set up? With the old Newtonian science, the concept of creation *ex nihilo* came to seem improbable or literally meaningless. Newton himself was a passionate believer in his fashion, even writing commentaries on the prophecies in the book of Daniel in his spare time. Like Aquinas, he believed that God had ordered everything and set the world in motion. Studying what God had actually done, as opposed to what some people erroneously believed he had done, was for Newton a form of piety. Nevertheless, the cultural impact of Newton's physics was to "rob the universe of its story." Where earlier biblical Creation accounts had given the universe a meaning and direction, Newtonian physics unintentionally suggested an eternal, mechanical, purposeless, and mostly lifeless structure. For many people who fear that religion may obstruct the benefits of science, defending that old Newtonian ideology is identical with defending science.

But with the advent of some modern scientific discoveries in this century, that is no longer the case. Two discoveries in particular, one at the macrocosmic level, the other at the micro level of particle physics, have separated us from what people took to be the old Newtonian view: relativity theory and quantum physics. Einstein's discovery that time is a dimension of reality like space returned even scientific thinking about cosmology to a point first articulated by Saint Augustine: that creation means precisely the emergence of time and space from the Creator, not an action occurring within a world that already exists, so that all the world — not just the evolution of stars, galaxies, and planets but the life forms with which we are familiar as well — bears the stamp of a unidirectional cosmic motion. When we turn, later in this chapter, to some recent discoveries about the so-called balance of nature, this cosmic directionality will show considerable relevance.

As science had developed prior to this century, the assertion that the world had been created and set in a particular direction had seemed less believable with each new discovery. God seemed, at most, the distant origin of something that otherwise was quite mechanical, the absentee landlord of creation, as some Deists began to think of him. The French mathematician and astronomer Pierre Laplace (1749-1827) theorized that if we could know the position and velocity of every particle in the universe, we could predict the future of all things. This clearly reduced the physical world to a closed materialistic system, perhaps eternal in itself, with no outside influences. When asked by Napoleon where God fit into this universe, Laplace answered famously: "Sire, I have no need of that hypothesis."

Life, at first, seemed to be an exception to mechanism and materialism. But soon every exemption from strict mechanics came under direct attack. Slightly earlier, the French physician and philosopher Julien La Mettrie had speculated that living beings, even humans, were nothing more than machines, *L'Homme machine* (Man a machine); in a notorious definition, La Mettrie described human beings as "perpendicularly crawling machines." Rapid developments in the knowledge of human anatomy and organ functions seemed to suggest, if not actually prove, that La Mettrie was right. Man, mind included, seemed explainable in terms of the mechanism found elsewhere in the universe.

The New Universe

The first decade of the twentieth century immensely shifted the scientific grounds for the old mechanistic universe and the human machine, so much so that we are still trying to come to grips with what it all means for our views of ourselves and our world. As scientists began to look into evidence about the origins of the universe, they considered various possibilities. The universe might be in a kind of steady state, as it mostly seemed in Newtonian physics; it might be expanding and, according to the second law of thermodynamics, headed for virtual extinction as matter became more and more thinly distributed through infinite space; it might expand and contract periodically; or it might be in a kind of continuous creation mode in which matter was being spontaneously generated somewhere to replace the galaxies that were moving further apart. What scientists found—now established so firmly that it is usually called the *standard model*—is the idea popularly know as the big bang, the original eruption that gave birth to the expanding and developing universe.[8]

Since relativity had established that time and space were part of each other, scientists did not labor under the illusion that such an event would be like a bomb going off in a vacuum. As hard as it is to visualize (as hard as Augustine's *creatio ex nihilo,* and for similar reasons), the generally accepted version of the big bang envisions time and space themselves emerging from what scientists call the initial "naked singularity." Emerging, in this context, must be understood to mean that not only matter and energy, as we are familiar with these in the everyday world, appeared with the big bang, but also time and space and the possibilities for matter and energy to act relative to one another according to certain laws. In 1990, scientists were first able to confirm, by means of the Cosmic Background Explorer (COBE) satellite, the approximately 3° K (i.e., approximately -572° Fahrenheit) cosmic background radiation from the initial fireball of the big bang, the figure predicted by computing initial energy and the current size of the universe.[9] The creation accounts and myths produced by the various human cultures had, for the first time, something roughly approximating a scientific confirmation.

Though some writers have rushed to claim the big bang as proof of creation, that is not at all certain. In fact, as early as 1951, when the big bang was far less established and far more pure theory, Pope Pius XII, in a speech to the Pontifical Academy of Sciences, reacted enthusiastically to the growing consensus within the scientific community with the pronouncement: "True science to an ever-increasing degree discovers God as though God were waiting behind each door opened by science."[10] Some scientists agreed; others took a strongly opposing view.[11] Caution seems warranted. The big bang is certainly a picture of the universe more easily reconcilable with the biblical view than was Laplace's or even Newton's. Time and the universe have an overriding direction and story again, as in the Bible (though like the biblical accounts, they are clearer when we look to origins than when we look to final ends). That story — of an unimaginable release of creative energy that resulted in everything we know, including ourselves and our science — orients us in significant ways within cosmic processes. But many explanations for its meaning other than biblical ones may prove plausible.

Still, it reminds us that the biosphere of earth continues to be influenced by cosmic processes; it is not only recurring cycles. So far as we can judge, life would never have appeared on earth if the planetesimals (themselves the debris of an early generation of stars) that came together into the terrestrial mass had not been spurred by later impacts into new and more complex chemical forms. Even the water that makes up so much of the planet's surface and gives it so many opportunities for life may have arrived mostly through melting comets at times when the earth was cool enough not to simply disperse it in evaporation.[12] The moon that seems so tranquil in the night sky to us, and plays so large a role in tides, was probably earlier broken off from Earth by the impact of a large asteroid. And it appears that some of the fluctuations in temperature that the earth experiences are correlated with perturbations in its orbit or sunspot activity.[13] Our biosphere is thoroughly rooted in ongoing cosmic forces at every level. For some, however, that history is merely a series of accidents. But now, scientific theory itself is beset by what seem fantastic and counterintuitive claims.

The Meaning of the Improbable

Paul Davies, a theoretical physicist, has given an account of various current cosmic theories in his book *God and the New Physics.* Davies has an interest in religion of a nonorthodox sort, and his book is arresting for the light it sheds on not only the counterintuitive mathematical world current science presents but also the religious hypotheses compatible with that world. In a few hundred pages Davies examines theories, rooted in older scientific assumptions, that our universe is the product of mere chance. The mathematics show that there is only an inconceivably minute probability that this is so, but it is not wholly ruled out by the new science. He also describes how parallel branching universes, mirror universes, perpetually expanding and contracting universes, and other outcomes based on other causes might also explain the process from the big bang to the world we experience.

Though Davies is not much concerned to attack or defend any position taken by any specific religious body, he concedes that the fundamental constants present early in the process of cosmic expansion seem to suggest that our universe was designed. There is a sensitive adjustment of these constants and the four basic forces in nature (gravity, electromagnetism, and the strong and weak nuclear forces) to one another, remarkably arranged out of almost infinite possible combinations. Scientists seem certain that all but an infinitesimal fraction of these vast possibilities would have resulted in a universe of black holes rather than galaxies and life. The summary Davies offers of the sensitive balance between the big bang and the restraining force of gravity is so striking that it bears repetition in full:

At the so-called Planck time (10^{-43} seconds) (which is the earliest moment at which the concept of space and time has meaning) the matching was accurate to a staggering one part in 10^{60}. That is to say, had the explosion differed in strength at the outset by only one part in 10^{60}, the universe we now perceive would not exist. To give some meaning to these numbers, suppose you wanted to fire a bullet at a one-inch target on the other side of the observable universe, twenty billion light years away. Your aim would have to be accurate to that same part in 10^{60}.[14]

The superscripts after each 10 indicate how many zeros need to be placed after the '1.' A negative superscript indicates the reciprocal of that number (1 over the number $10^{-3} = 1/1000$, or 0.001).

Roger Penrose, another scientist turned popular writer, has put the improbability of our universe even more remarkably:

> This now tells us how precise the Creator's aim must have been, namely to an accuracy of one part in $10^{10^{123}}$. This is an extraordinary figure. One could not possibly even *write the number down* in full in the ordinary denary notation: it would be '1' followed by $10^{10^{123}}$ successive '0's. Even if we were to write a '0' on each separate proton and on each separate neutron in the entire universe — and we could throw in all the other particles as well for good measure — we should fall far short of writing down the figure needed.[15]

If these immense indications of the natural improbability of our universe do not prove that we are the product of cosmic design, they certainly weight the case heavily in design's favor. Some theorists have proposed an anthropic principle, the notion that the forms we ourselves bear must have been built into those early constants. The notion seems fantastic. But as Stephen Hawking, the greatest living cosmologist and no easy ally of religion, has stated: "The odds against a universe like ours emerging out of something like the Big Bang are enormous. . . . I think there are clearly religious implications wherever you start to discuss the origins of the universe. There must be religious overtones. But I think most scientists prefer to shy away from the religious side of it."[16]

The Hawking Phenomenon

It is worth pausing briefly on Hawking because he has become for the second half of the twentieth century, in fame if not securely yet in fact, what Einstein was to the first half. Hawking was appointed Lucasian Professor at Cambridge University in 1979, just over three hundred years after Isaac Newton had occupied the same post. He was only thirty-seven years old, and the appointment was a tribute to the great work he had already achieved in physics and cosmology. But Hawking also became a media presence, primarily because he presented such an amazing human image. In his twenties, he had contracted amyotrophic lateral sclerosis (ALS), better known as Lou Gehrig's disease, and over the course of a decade had lost almost entirely his ability to move and speak. Hawking, a shrunken, childlike figure in a wheelchair, first created a computer setup that allowed him to communicate and went on to use it to publish breathtaking new theories about the universe. Possibly more than Einstein in the Bern patent office, Hawking represents the astonishing capacity of the human mind to encompass worlds. His work became known through television specials. His 1988 book, *A Brief History of Time,* became a best-seller, even though the argument in that short work was all but impenetrable even to intelligent nonscientist readers. Its concluding sentence that we might someday "know the mind of God" seemed to clash with his stated opinions elsewhere.

Hawking continued working at theories that may not, strictly speaking, require a cause like a Creator to get everything started or set them on the course they took — a view that seems an ingrained characteristic of science strictly as science. In almost a perfect parallel to the papal controversy in 1951, Hawking and John Paul II had a run-in at a 1981 meeting of the Pontifical Academy of Sciences, where the pope told the audience:

> Any scientific hypothesis on the origin of the world, such as that of the primeval atom from which the whole of the physical world derived, leaves open the problem concerning the beginning of the Universe. Science cannot by itself resolve such a question; what is needed is that human knowledge that rises above physics and astrophysics which is called metaphysics; it needs above all the knowledge that comes from the revelation of God.[17]

Earlier at the same conference, Hawking had presented his "no boundary" theorem, which argues that there is nothing outside of time and space, even though the universe is finite and has a measurable time of existence. Attendees sympathetic to the pope were distressed when the Holy Father knelt to spend a long time in conversation with Hawking in his wheelchair. But whatever transpired between them, Hawking seems to have identified himself now with a view of the universe that could dispense with the need for a Creator. Yet a note of doubt creeps into his complex theory as he asks an age-old question: "What is it that breathes fire into the equations and makes a universe for them to describe? The usual approach of science in constructing a mathematical model cannot answer the question of why there should be a universe for the model to describe. Why does the universe go to all the bother of existing?"[18] As John Paul II rightly told the audience where Hawking sat, that kind of question goes beyond physics to metaphysics.

The Quantum World

The other major breakthrough in science this century, quantum theory, presents an even more remarkable and, in many ways, troubling view of the fundamental structure of matter.[19] Up until just about the time Einstein was arriving at his special theory of relativity, the mainstream view of the atom was that it was something like a miniature Newtonian solar system: electrons circled around the nucleus much as planets travel around the sun. Energy and matter, as Einstein later proved, were convertible into each other, but basically matter in motion exhausted reality. However, as in relativity theory, light posed a special problem. Briefly, light behaves like a particle in some instances and a wave in others. In other words, it's as if the ocean could be described at times as the droplets of water that make it up and at other times as the phenomenon of waves striking the shore. To a commonsense view, in our world this is impossible: water is not the same as wave motion, though water may be made to move in waves. Quantum theory confounds that and many other common-sense notions in saying that at certain times the matter itself *is* a wave and, even more remarkably, that whether it is a wave or a particle depends on what we, as observers, ask about it.

For the other massive notion that quantum theory introduced is that observation affects reality. In an almost exact contradiction to Laplace's understanding that given enough knowledge we could predict the future course of the entire universe, Werner Heisenberg discovered that at subatomic levels we cannot know both the position of a particle and its velocity. We can know only one or the other. Moreover, this incapacity does not stem merely from the current limits of our measuring devices; Heisenberg's uncertainty principle seems to demand that such indeterminacy is part of the very structure of the physical world itself. Though the consequences of such a theory are complex scientifically as well as philosophically, the whole discovery suggests that the old view of the split between matter and mind, evident in philosophy since Descartes, is, at the very least, changing and soon may even be put on a new basis.

The observation of particles, according to quantum theories, itself changes them. How this can be further confounds normal understanding. Especially if we take the usual view today that mind is merely a kind of epiphenomenon built up from the chemical processes in the brain, it is difficult to see how those processes could affect objects in the physical universe. Yet physicists say, and have the data to prove, that after certain events at the subatomic level, the result does not take definite form until someone actually observes it, whereupon it "condenses" into actuality from an uncertain state. Even more astonishing, some of these effects seem to operate faster than the speed of light—an impossibility according to relativity theory—or perhaps even to operate backwards in time, later observation changing previous quantum events. What special power can an observer have over matter that would produce this effect, particularly if the observer is a priori regarded, not as a special creation of God, but as the product, albeit complicated, of the very interactions the observation is said to fix? We do not know, but quantum theory has proved its truthfulness as well as its utility in making possible the development of instruments like lasers.

Quantum theory is a much more complicated business than has been sketched here. Some theorists have suggested that the universe came into being to produce observers like ourselves with our strange property of enabling indeterminate quantum states to become fixed. In the strange way that the big bang and quantum theory interact, in some descriptions every place in the universe was originally the point at which the process began and retains some similarities to it. Scientists sometimes take this a step further and argue that our present attempt to measure and observe quantum events at the beginning of time actually involves us in some weird way in the coming-to-be of our universe. To the dismay of some scientists, relativity and quantum theory do not fit harmoniously together, either. The presence of faster-than-light effects in quantum mechanics, an absolute impossibility in relativity theory, for example, has led some scientists to speculate that neither can be exactly right. And a whole new set of theories involving "strings" and "superstrings" is being elaborated to try to reconcile apparently contradictory elements of the universe into a "theory of everything."

Some writers, such as Fritjof Capra in *The Tao of Physics* and Gary Zukav in *The Dancing Wu Li Masters,* and David Bohm from a slightly different perspective, have claimed a religious significance for quantum theory. (Bohm elaborated a theory of "implicate order" and a "pilot-wave" interpretation of quantum theory that sought to remove some of its apparently unbelievable consequences.) They see it as akin to paradoxical religious practices like Zen Buddhism. Sensitive writers like J. C. Polkinghorne who have training in both Christian theology and physics have argued that such attempts depend "too greatly on purely verbal parallels to be convincing"[20] and have made contrary arguments showing the theory's compatibility with Christianity.[21] No less distinguished a scientist than Stephen Hawking, at least at an early phase in his career, has called the attempt to link physics and Eastern mysticism "rubbish," saying, "The universe of Eastern mysticism is illusion. . . . A physicist who attempts to link it to his own work has abandoned physics."[22]

New Sciences of Life?

The physics we have looked at up to this point forms the background for any intelligent current discussion of life and the biological sciences, which are both topics central to the discussion of environmentalism. Despite all the ferment in physics, biology often seems wedded to older scientific notions. Biology, of course, was deeply revolutionized in the nineteenth century by Darwin, who argued that the emergence of life, including the appearance of man, could be accounted for through a series of minute changes in response to the environment or, as he formulated it, "natural selection through random variation." Given the scientific record as it has now been indisputably established, there is no question about the successive appearance of certain life forms on earth. Indeed, it fits entirely with the way we now know that the universe itself evolved from simpler forms of matter, through a first generation of stars, into the universe of heavier elements and proliferation of life on earth with which we are familiar. The question, however, is whether mere randomness, rather than some form of purposeful design, produced life. As we have seen in the case of cosmology, the question of accident/design seems weighted towards design by the precision of the four forces in the universe. But in biology the case is far more difficult to resolve, because life is much more complicated than inanimate matter, and the interaction of physical processes and life forms on earth presents multiple and irreducible complexities.

Even before we arrive at the question of man — with the seemingly unique features we display of thought, will, and conscience — exactly how life has emerged is in hot dispute. The most vigorous recent defense of classical evolution theory, Daniel C. Dennett's *Darwin's Dangerous Idea*, reformulates it as a competition between those who believe in "skyhooks" (mysterious lifting devices or principles that would explain the relatively rapid development of life compared with the mathematical probabilities of random chemical reactions on earth) and those who believe in "cranes" (that is, devices that depend on stories already built below to add on additional levels). Both cases try to account for the seemingly impossible speed with which life developed if we allow only for random variation. Dennett's innovation is to posit the emergence at some point of "memes," or transmissible information, something like ideas. Once the human mind appeared with its capacity to manipulate such memes, the mind became a "crane" capable of very rapid cultural and scientific evolution.[23]

But memes seem to solve the issue no more than molecules. As Dean Overman has demonstrated persuasively in his brilliant *A Case against Accident and Self-Organization*, the improbabilities that make the universe appear designed apply to the emergence of life as well.[24]

Life appears to have arisen on earth in a mere 130 million years, short by cosmic standards, from the time the early earth was still inhospitable to life until the first single-celled organisms. Life already assumes the stunning improbability of a universe that Overman terms "compossible" with life. But the sheer chances against life itself are staggering. Life as we know it requires complex chemical compounds known as enzymes. These are constructed from amino acids in complex sequences and help to assemble proteins. To get even the simplest paramecium, you would need around two thousand such enzymes. Scientists inclined to the self-organizing properties of matter have calculated that the chance of randomly achieving even that simple life form is something like 1 in $10^{40,000}$.

Furthermore, Overman notes, scientists have been confusing order and complexity. It is fairly simple to account for how order in crystals and other regular natural products arose. But complexity, like that revealed in the human brain or a DNA molecule, requires far more information than is available in the chance physical and chemical processes that Dennett believes happened, even if these were allowed to go on for the age of the known universe. Thus, all living things seem to exceed by wide margins the kinds of organization available from a merely materialist universe. In fact, they start to take on the kind of meaning and directedness we are familiar with in systems like language, which is to say that they almost look as if they were "spoken" intentionally by a mind. In religious terms, which Overman does not invoke, life seems to participate in a kind of *logos,* a Greek word adopted by theologians to represent the "Word," or reason, that God spoke when he created.

The very improbabilities that science itself now sees in Darwinian theory, however, have given rise to another school of biological interpretation. While still recognizing that life emerged in stages over long periods (no return to biblical-scale creationism), some biologists have begun to examine what appear to be unresolvable dilemmas for the emergence of life by sheer chance, particularly at the molecular level. Michael J. Behe, a professor of biochemistry, has presented one of the strongest recent cases showing that it is difficult if not impossible to conceptualize how some of the biochemical reactions resulted in complicated structures even for one-celled organisms.[25] The current defenders of Darwinism, he says, not only Daniel Dennett but other distinguished figures such as Richard Dawkins, Elliott Sober, and Michael Ruse, have tried to propose various schemes by which the variations generated are random but the selection process is not. In this view, the environment, with almost a kind of prescience, eliminates all "unsuitable" variations.

This, however, says Behe, begs several questions and ultimately will not do. The development of a swimming appendage known as a *flagellum* in a paramecium, for example, would require that one set of chemical reactions "know" that a parallel set will occur in order for the first reaction to have an evolutionary survival value. This attributes the kind of intelligence and choice that we associate with persons to the physical and chemical processes of nature. Behe demonstrates rather convincingly that even given long cosmic times, such parallel developments appear not only improbable but inexplicable. His survey of the biochemical literature found not a single article that proposes to explain even the simplest of such processes—after over a century of scientific acceptance of Darwinian theory.

Obviously, if the same biochemical test is applied to a complex assemblage of cells like a horse or, even more, the human body, we are faced with the image of billions of reciprocally tuned impossible emergences. DNA itself is such a complex molecule that it is difficult to see how it could have emerged from such a process. One of the discoverers of DNA, Francis Crick, has even been driven to the hypothesis that perhaps life was deliberately planted here by extraterrestrial beings.[26] Crick's proposal reflects both his perception of the unlikelihood that the DNA molecule developed by wholly natural processes on earth over the past five billion years and colossal question-begging: if DNA could not have developed naturally here, how did it develop elsewhere for the extraterrestrials to "seed" our planet?

The Play of Life

What, if any, importance does this new biology have for questions about man's place in the environment? Some of the traditional religious beliefs are still gone forever. The earth is far older than the several thousand years represented in the Bible, and the universe is older still. Humans, animals, plants—life, in short—are still not properly explained by evolutionary theory or any other science, but without question they are not beyond some further explanation, at least so far as their physical processes are concerned, by stronger theories. Still, one of the striking things absent from all that we have mentioned thus far is the kind of static balance that most environmentalists and religious people assume is the basic character of life on earth. We have been talking about a developmental direction for both the universe and life. Some balances do exist over long periods of geologic time, but these have to be situated within another motion that does not allow nature ever to remain in simple balance. As two distinguished geologists have observed: "Perhaps the most important lesson to be learned from the history of Earth is that our situation in life is not static, and that we (personally and as a species) have to be on the lookout for new environmental niches. . . . There never has been a 'fixed reality' for Earth."[27]

And even if we are able to use harmonies in nature for some explanatory purposes, what are we to say about that massive anomaly, the human mind? Human consciousness remains the most mysterious phenomenon in an already quite mysterious universe. And all attempts to crack that secret, even some relatively ingenious recent ones suggesting solutions in combinations of quantum theory and relativity theory, seem pitiful failures. However science may eventually come to regard this question — whether the Ultimate Skyhook, God, seems required or whether Cranes internal to nature are sufficient — the whole question of the human mind draws us back towards the environmental question. From the typical environmental perspective, the human mind, even at its most benign, is an accident waiting to happen. It is not closely tethered to its environment the way animal responses to stimuli generally are. It is restless and impatient, given to flights of fantasy and unnatural longings, stubbornly attached to particular offspring and communities, and not readily willing to submit itself to some order supposedly imposed by nature and nature's God, whether in the moral order or the physical.

Intelligence as Adaptation

But if we step back from these immediate concerns, a larger question arises: how could this allegedly ill-adjusted creature have arisen from the natural order? If evolutionary explanations are in large measure true, what was the evolutionary value of this flighty, somewhat unstable creature? Not surprisingly, some neurophysiologists have begun to think that there is an evolutionary, adaptive value in the kind of minds we have. And they do so precisely because hominids, in the 2 million or so years that they have been around, have had to deal with sudden, naturally occurring, environmental changes that made a jack-of-all-trades creature better fitted to survival. In the introduction to the present volume, we mentioned some of the instabilities in nature that explode any simple notion of a fixed and benign natural order. But if the neurophysiologists are right, those very sudden swings in natural environment may have been the conditions that led to the sudden explosion, by geologic standards, of human intelligence to deal with them. Far from being a kind of fast-moving and reckless alien intruder, then, human innovation and creativity may be rooted in that whole fifteen-billion-year cosmic process that we have been examining here.

In fact, some environmental analysts believe that most living things on earth, themselves survivors of naturally changing conditions and even catastrophic events, have probably developed some similar adaptability to environmental change. Gregg Easterbrook has summed up what we misperceive to have been a tranquil history of life prior to the advent of the human race:

> The living environment of Earth has survived ice ages; bombardment of cosmic radiation more deadly than atomic fallout; solar radiation more powerful than the worst-case projection for ozone depletion; thousand-year periods of intense volcanism releasing global air pollution far worse than that made by any factory; reversals of the planet's magnetic poles; the rearrangement of continents; transformation of plains into mountain ranges and of seas into plains; fluctuations of ocean currents and the jet stream; 300-foot vacillations in sea levels; shortening and lengthening of the seasons caused by shifts in the planetary axis; collision of asteroids and comets bearing far more force than man's nuclear arsenals; and the years without summer that followed these impacts.[28]

Easterbrook is not suggesting — and neither should anyone else — that these things that nature does to itself should make us careless of what we do to nature. At times, we intervene massively and quickly in ways that are not good for nature or ourselves. But anyone who ignores the cosmic dynamic in the perturbations of our planet will not only mistake the nature of God's world, he will misjudge what is and is not of serious environmental concern.

If true, this explanation turns the whole environmental question in a different direction. In an older view — one supported both by unexamined ideas of harmony in the Bible and by the gradualist theories of the early modern geologists and biologists — environment was thought of as shifting only incrementally. There might be local aberrations that disturbed balance, but the physical world and the creatures that had come to occupy various environmental niches were thought to be in a steady relationship with each other. Indeed, the notion of the earth as Gaia (the name of a Greek goddess that James Lovelock and Lynn Margulis first used to hypothesize that the entire earth was like a large self-regulating cell) became popular because it seemed to encapsulate all the older views in an environmentally potent form.[29] Lovelock himself later became skeptical of some of the uses to which the idea was put and turned to studying ocean life, which he thought more significant. But the image remains in environmental literature that animals and their habitats exist in stable equilibrium, plant life growing complexly until it reaches a climax condition. And these habitats are coordinated with one another, if not by an earth goddess then by some self-regulating mechanism.

The only problem with this picture is that, except for restricted conditions, it is not true. It is true and useful when we try to think about discrete habitats and species. If you want, say, to preserve the spotted owl in the Pacific Northwest, you need a certain type of habitat of a certain size. Different habitats do interact with one another and achieve a kind of stability over varying periods. And promoting that kind of stability, along with protecting creatures, is what many people hope environmentalism will be able to do. But the science of the case is not simple, and it does not always support the common view. The relationship of habitat to preserving species is fairly clear. But there are irreducibly complex systems on earth that will never admit of prediction: weather further than a few days into the future is one such system, and climate may be another. (In a parallel to the way quantum physics has defeated the old Laplacean view of perfect predictability, chaos theory applied to the biosphere suggests that the past does not, strictly speaking, control the present, or the present the future — a truth that opens up possibilities for innovation and the emergence on Earth of truly new things.) Anyone who has ever looked closely at the scientific debates over climate change, for instance, knows that the number of factors involved and their complex interactions make any sure predictions virtually impossible.

Natural Imbalance

As we have seen, the earth has been bombarded with stray cosmic junk, which helped and continues to help move geologic and biological processes along. For much of earth's history, those bombardments added to the world's potential, both organic and inorganic. After life came along, destruction often preceded new creation. Around 500 million years ago, something happened—no one knows exactly what — that allowed what geologists call the Cambrian explosion of new forms of life. About 250 million years ago, something else, probably the impact of a large asteroid or comet, destroyed 90 per cent of existing species. But that cleared the way for new kinds of life. Something like 65 million years ago, another strike, probably in the Caribbean, killed off the dinosaurs and set terrestrial life on a different course. In the two million years that humanlike animals have existed, the world has been mostly in ice ages.

And there are even more troubling dimensions to these sudden changes. During the last ice age, around 73,000 years ago, when the environment was already quite hostile to emerging human life, a massive volcanic eruption in Sumatra caused a centuries-long drop of from 10 to perhaps more than 25 degrees Fahrenheit in average global temperature. In rough terms, this was like adding another ice age on top of the ice age that already existed. Scientists have also uncovered evidence that 65,000 years ago, owing to harsh conditions naturally produced, the human population declined to as few as 10,000, living in Africa.[30] When we add up all the bad things we have done to nature that have caused, and may yet cause, massive changes in natural patterns, it is good to keep in mind how precarious the survival of the human species has been under purely natural conditions.

This is a troubling vista if you assume that the Creator must work through what we think of as gentle benevolence. Some theologians, seeing that vista and aware of what continues to go on in nature, have theorized that God cannot be all powerful, though he is still all loving. Of course, this reintroduces the notion, expelled by the biblical notion of creation, that some evil element intrinsic to the world is mixed up with the Creation. But that is a drastic remedy for the indisputable facts in the case. It may be that, if we give ourselves some time to digest the geological and biological facts, as with cosmology we may find the new world presented to us by science more compatible with Creation theology than may first appear.

Geology and the Bible

For example, the workings of God in the cosmos are no stranger than some of the events we read about in the Bible, if we look at the Bible stories without preconceptions. Nature in general changes slowly over long geological periods, but not always. For example, there was a time when the biblical story about Noah and the Flood was regarded as a suggestive but fictional event dreamed up by inhabitants of the Middle East. Similarities between the biblical account and Babylonian myths, however, troubled many scholars, who began to look further. We now know from incontrovertible geological sedimentation that a flood of some kind occurred in the Middle East around 4000 B.C. How exactly it was caused and how long it lasted are matters of dispute.[31] Today, it is common when, say, the Mississippi River floods, to assume that maybe human action has altered nature's more gentle paths. But sudden, surprising turns in nature are not as rare as we might think and have left their effects in human records like the Bible.[32]

On our continent, Native American legends of great floods also exist. One in particular seems to have good geological foundation and may help explain similar so-called myths. The southwestern corner of what is now the state of Washington consists of "channeled scablands" that were formed after the last ice age. The theory is that, as ice melted and large inland lakes were formed, some of them were dammed up by relatively narrow tongues of ice. When those tongues melted sufficiently, the dams suddenly burst, sending down huge and powerful cascades of water, hundreds of feet high, that reconfigured immense landscapes. There may have been thousands of such events, if on a smaller scale, in North America within human memory. It is worth reminding ourselves, as a distinguished researcher into these phenomena has remarked, that far from being tranquil and benign, nature in this one area alone is quite active: "floods, plural, are a universal geologic phenomenon."[33]

Similarly, some of the events associated with the Exodus account in the Bible may have explanations that are, in the most immediate sense, natural. Scientists speculate that the sudden disappearance, around 1500 B.C., of the great Minoan civilization on Crete may have been caused, in part, by a huge volcanic eruption on the nearby island of Santorin. (The legend of the lost island of Atlantis may reflect a distant memory of these events also.) Both the Bible stories and Egyptian papyri from the period speak of the sun being veiled, rivers turning the color of blood, plants withering and drying up. All these things are consistent with the well-known effects of large volcanic eruptions. We are not certain about the date of the Exodus, but the First Book of Kings says that Solomon started building his temple 480 years after the Hebrews left Egypt. If we add that number to 970 B.C., the most likely date for the start of Solomon's reign, we get 1450 B.C., a date quite close to the Santorin eruption.[34]

Data like these about floods and extraordinary events used to be used to debunk supernatural causes of things narrated in the Bible. But for anyone who believes that God's providence can operate through natural causes as well as special interventions in nature, these data should more readily be viewed as confirming the basic truthfulness of the improbable things the Bible records. Furthermore, they also remind us of the main point we have been tracing here: that, whatever steps we may wish to take to reduce human damage to nature, nature itself and the God who created it show little in common with today's usual environmental view of the world as a constant, benign, and nurturing place except when foolish or outright evil human acts disrupt it.

The New Ecology

Daniel Botkin, an ecological scientist and adviser to several international bodies, in his book *Discordant Harmonies* summarizes much recent research on the "balance of nature." He cautions that we have to become more accustomed to thinking of "the dynamic rather than the static properties of the Earth and its life-support system, and the acceptance of a global view of life on Earth."[35] But for him, thinking globally has a different meaning than for most who use the term. The biosphere of earth shows intrinsic and natural change "at many scales of time and space," and not recognizing that fact has led many concerned with the environment "to emphasize the benefits of doing nothing and assuming that nature will know best."[36] But nature from a human perspective, and even from the perspective of those who would like to preserve the types of nature with which they are familiar, does not always know best, and Botkin calls on us to build a great civilization that will manage and enhance the natural world.

The balance that many believe exists in nature Botkin prefers to call a discordant harmony, "at some times harsh and at some times pleasing." For example, it used to be thought that predators and prey in undisturbed habitats, or competing plant species, reached a "natural" balance. Closer looks at what actually occurs reveal that there are blooms and crashes among these opposing forces that are not easily explicable. Outbursts of locusts and other insects suggest that many species populations vary widely in nature. And even attempts to confirm that predators and prey reach a rough balance are difficult: "there is little evidence that the result is a constancy of nature, a perfect balance in the classical sense."[37] Sometimes, even in the absence of predators, natural animal populations swell and decline.

Forests, too, change rather quickly owing to variations in the overall biosphere. Indeed, if we take the larger geological time scales, the relative stability of the past few thousand years — the period when human beings have left historical records and had formative experiences — is rather abnormal: "The old idea of a static landscape, like a single musical chord sounded forever, must be abandoned, for such a landscape never existed except in our imagination."[38] Even tropical areas like Africa have showed wide variations. An African expert, David Livingstone, has written that "the African environment is capricious, not stable, and apparently has been so for at least several million years."[39] And Andrew Brennan has added that the biodiversity we all cherish is not necessarily the result of forests remaining undisturbed: "natural forests sometimes display diversity precisely because they are disturbed by factors such as fire, weather damage, and human manipulation."[40]

The resiliency of forests has had striking recent confirmation: only three years after many square miles of woods were destroyed by the eruption of Mt. St. Helen's volcano, the area visibly began to come alive again.[41] Many of the wilderness areas we think of as pristine—the Amazon, for instance—are actually the result of a long interaction between human beings and nature.[42] That interaction may, of course, be benign or harmful, depending on what we humans actually do, and in recent years far too much of the Amazon has been burnt far too fast. But to preserve the forests with which we are familiar from natural variations would be a monumental task. Botkin argues:

> If we want to keep some of these stages frozen in their current conditions, we can change the soils by turning them over or adding fertilizers. That is, we can substitute our own energy, time, and resources to replace nature's natural processes. We can do this in some cases for a long time and in all cases for a short time, but we cannot freeze all of nature indefinitely in a single state.[43]

Or at least that is the case in the current state of our knowledge. Most environmentalists would probably object to the very notion that we should take charge of nature in the way Botkin indicates might be possible. But the choice seems to be either to do that or to allow nature to take its course, a policy that we may not find has happy outcomes.

A Changing Creation?

Botkin directly confronts the view, common in both classical thought and biblical interpretation, that there must be a natural harmony. All things were supposed to be fitted and well ordered to their place in the Great Chain of Being. The usual explanation when we find that they are not is that we must have disturbed the order by what we have done or failed to do. Like Newton, however, Botkin thinks we have let what some people think about nature confuse us about what nature really is. We have to go to the real science to see, as the biblical book of Job most emphatically shows, that we do not understand all divine purposes in making the world the way it is. Our wonder at nature is a proper response to a remarkably complex and beautiful system. But empirical evidence indicates that we may need to take, with our new powers, a far more active role to arrive at what we believe nature should be: "The idea of a divinely ordered universe that is perfectly structured for life has persisted, if often beneath the surface, influencing the development in an obscured way of the interpretation of the environment, nature, and the role of human beings in nature in our time."[44] Some of those assumptions need to be brought to the surface and submitted to scrutiny the way earlier notions of the cosmos have been by real science.

That examination might lead us to a view Botkin does not mention: that we may have wrongly believed that creation, because it is good, is therefore as perfectly good as God, something the orthodox tradition would repudiate. There is much to learn from natural forms and much that we have overlooked in nature in our past three hundred years of practical progress, but the notion of nature as our Mother, as a being greater and better than ourselves is — and always was — mistaken. Nature is our sister, a fellow creature, and one that, like all creatures, may need reform. Botkin gives several examples of how organic, rather than mere civil-engineering, techniques can better shape nature. A forest, for example, passes through a series of stages, reproduced in most forests, but one-way for each particular forest. Some of our responses to change need to acknowledge that; others need to take into account complexity and nonlinear processes. But the same principle remains behind all approaches: nature has no single end in itself. "People are forced to choose the kind of environment they want, and a 'desirable' environment may be one that people have altered."[45]

Of course, we will want to continue reducing the noxious effects some human activities have on the environment on which we depend, and we need to observe nature carefully to know what effect we are having. Large steps have already been taken in that direction, and air, water, and soil today are far better off as a result. But if the environment is capable of rapid changes, whether natural or man-made, it may be that nature or God built special capacities into us to enable us and other living beings to cope with, and perhaps even prevent, such events. Though over the last century and a half we damaged nature badly through uncontrolled industrial development, nature too seems to have remarkable adaptive qualities and resilience built into it, perhaps from its long interaction with sudden geologic threats. For instance, Rachel Carson, the fountainhead of the modern environmental movement, predicted that pesticide use would cause forty species of birds to go extinct by now. In fact, because of the human capacity to adapt—which is to say intelligent and successful management of a humanly induced environmental problem—thirty-three of those species are stable or growing today; only seven are down.[46] Not a perfect response, but not the end of the world either. In any event, we can see that over the past two or three decades our steps to reduce human impact on nature have led to quick rebounds in rivers and lakes, even around industrialized areas. Wilderness returns quickly after natural or man-made disasters; Prince William Sound, once thought finished after the 1989 Exxon Valdez spill, is already back near its pre-spill condition. Both we and nature have remarkable flexibilities and resiliences built-in.

Actively Managing Nature

And we may need them rather soon. In a remarkable essay entitled "The Great Climate Flip-Flop," William H. Calvin, a neurophysiologist, sorts out some of the challenges we may face in the not-so-distant future. Calvin notes that our current warm cycle began about 15,000 years ago as the last ice age ended. Long ice ages, on the order of 100,000 years, have been alternating with shorter, warm interglacial periods for millions of years. But he warns against seeing these as merely gradual, long-term cycles: "We now know that there's nothing 'glacially slow' about temperature change: superimposed on the gradual, long-term cycle have been dozens of abrupt warmings and coolings that lasted only centuries."[47] In the scientific literature, these swift natural variations are well documented. We have already seen how the Younger Dryas, a sudden period of cold after the end of the last ice age, plunged the world back into serious low temperatures. During that time, the forests of what is now Germany went through a period when they died out and looked like the scrubland of modern Siberia, says Calvin. He worries that we are experiencing a rapid, human-created warming through carbon dioxide emissions that may lead to a subsequent cooling.

The data about global warming are mixed. Anyone, for example, may call up the NASA satellite records from the Internet: NASA claims the satellite data show a slow cooling trend, while surface data show a sight warming trend. And the natural variations that lie behind what we may or may not be seeing further cloud analysis. During the last two thousand years, we have had periods of exceptional warm and exceptionally cool weather that have lasted for centuries. But quick and large changes also seem to be programmed into the earth's various systems. As Maureen Raymo, a researcher at MIT, has noticed, we have evidence that climate has changed rapidly in the past 1.5 million years, even before any human effect could have been implicated. "These climate swings are so dramatic that if we lived through one today, it would be like New England taking on Miami-like weather within a 25-year period"[48] — changes, of course, that dwarf even the worst-case scenarios for anthropogenic global warming.

Natural variations are worrisome enough, and recent human factors, like greenhouse gas emissions, Calvin thinks, may play into forces that warm the earth in the short run but may provoke disastrous ice age–like *cooling* in only a sightly longer run. One possible mechanism might be this. We know that the North Atlantic Current carries warm water from the equatorial Atlantic up towards the north, keeping England, Ireland, and the rest of Europe warmer, by North American standards, than their latitudes would normally suggest. If this current is interrupted—as seems to have happened many times in the past—much of Europe and other parts of the North Atlantic could become quite cold, threatening large sectors of their populations through damage to agriculture. In a world with nuclear weapons, it might also lead to ugly wars for resources.

The current is switched off, it appears, when the saltiness of the water that arrives from the south is diluted by too much rain or by sudden outflows from dammed-up fjords. (Less salty water will not sink and circulate back towards the south as it does at present.) Excessive rain in the north might be one consequence of higher global temperatures and, therefore, cause greater dilution. Depending on when and how often rain falls, it can have long-term effects. Whether the temperature rise occurs naturally or because of human activity, the result is the same. Another possible mechanism for shutting down the current is a sudden outflow of fresh water once trapped in a fjord by an ice dam. Calvin recalls that the Hubbard glacier in Alaska closed up the Russell fjord in May 1986, creating a natural fresh water lake eight stories high. We have already seen how the melting of such natural dams may have produced floods in various parts of the world and almost certainly created the vast channeled scablands in Washington. When such large water outflows dilute the salty water of the Atlantic around Northern Europe, they may have the same effect as the high rainfall.

What might either of these scenarios suggest for us? For Calvin, who fears man-made global warming and advises prudent steps to lessen our emissions of greenhouse gases, there is no escaping an active human role in managing nature. Another period of cold, which is due shortly if the ice-age rhythms of recent eras continue, would ravage the human race and its civilization. Calvin has no hesitation in saying that we need to use the highly agile minds that sudden environmental change may have given us to prevent that cycle from repeating itself to our detriment. So he proposes that, should we see the North Atlantic Current being affected, we could take the relatively low-tech approach of blowing up fjord ice-bridges, if they form, or diverting their waters to less sensitive areas. If warming begins to cause more precipitation, we may need to seed clouds so that the rain falls elsewhere or even use bargeloads of "evaporation-enhancing surfactants" in the sensitive areas. Whatever steps may be needed, Calvin identifies a common point:

> We cannot avoid trouble by merely cutting down on our present warming trend, though that's an excellent place to start. Paleoclimatic records reveal that any notion we have had that the climate will remain the same unless pollution changes it is wishful thinking. Judging from the duration of the last warm period, we are probably near the end of the current one. Our goal must be to stabilize the climate in its favorable mode. . . . We have to discover what has made the climate of the past 8,000 years relatively stable, and then figure out how to prop it up.[49]

This view, of course, cuts directly against the assumptions of most religious people and environmentalists. It is not hard to imagine the reaction to such a proposal: who are we to mess with God's created order, and what guarantees do we have that trying to stabilize it will not make matters much worse?

No Guarantees

The answer to the first question is that, whoever we are, no other being is likely to preserve what we wish to preserve. The earth will not much mind if human societies sharply decline or wholly disappear. All but a fraction of 1 per cent of earth's history took place without human beings, and presumably the earth and some forms of life can continue on their merry way without us. Whether this is a desirable outcome in human terms and whether it reflects responsible stewardship in religious terms are questions that each of us will have to answer. For most of us, however, a world without human beings and vigorous, advanced human societies would not be much of a world.

As for guarantees, we have none. The older wisdom of the religious tradition emphasized that we are all wayfarers here and that none of us has any assurance that he will even live out the day. The same is true for the race and the earth itself. Large objects like the one that killed off the dinosaurs arrive periodically. Indeed, certain elements in the biblical tradition warn us to prepare for an end to the world. Without falling into baseless speculations about the apocalypse, of which no man knows the time, it still may be that God put us here with the capacities we have precisely so that we can overcome the limitations of geologic forces and preserve a healthy and verdant nature into the bargain. To take this path would mean that we would also have to change our view of stewardship to encompass not only maintenance of an existing order but also vigorous intervention to prevent things from taking their natural and, often enough, undesirable course. We do not know whether we will have the wisdom and knowledge to do what we will need to do to preserve ourselves and the planet in a state conducive to human and natural flourishing. But our choice is not between reduction of activity and harm to the earth. We have been through a couple of centuries in which our knowledge and power grew enormously without our noticing some of its consequences. Fortunately, we have the luxury now of technologies that already exist, and more to be discovered in the future, that will allow both vibrant societies and economies and decreasing impact on nature. The developed world, with a few ongoing exceptions — most notably possible global warming — has already largely succeeded in this task, with more successes to follow.

Yet we should not be deluded into thinking we can create some perfect natural balance that nature itself does not possess. There will be failures and mistakes. But we have to choose: what kind of environment do we want? As in most human things, there are different answers to that question. Some of us in America would like to preserve more of the continent in the form it was in, or we imagine it to have been in, prior to the great European settling; others might choose a state more like colonial America. Along the East Coast, many might prefer a landscape humanized, as it is in parts of Italy, by a soft adaptation of the land to human presence. And still others will be content to live in large urban and suburban centers made more livable by parks, wildlife trails, deliberately zoned-off green areas. All of these forms, and many others, can be environmentally responsible ways of living, but they all have one thing in common: Since nature itself can probably survive and flourish in various ways, it is we, as the only forms of life that have the power to plan and foresee some of the consequences of our actions, who will have to decide about the future and the trade-offs for us and nature that human existence inescapably demands.

PART TWO

Some Case Studies

A Hopeful Interlude

The figures examined in the following chapters are not necessarily the best thinkers about religion and the environment. They have been chosen because, in one way or another, they represent influential or noteworthy positions in the current debate. Religious environmentalism arises from a variety of viewpoints: historical analyses, poetic intuitions, theological speculation, feminist thought, social justice concerns, and many others. It could hardly be any different, since to speak of God and nature is to address virtually everything. This section cannot claim to be exhaustive, but it provides a broad-brush survey of almost all the main currents in religious environmentalism. Everyone who thinks about environmentalism hopes that both human beings and the planet will be better off in the future. The brief survey of the conflicting environmental claims in this interlude is a personal reading of the literature that concludes we have good reasons for hope. To address even one of these subjects properly requires an expertise that I do not possess. But we are fortunate in having a wealth of good empirical studies available to us today. The reader is encouraged to make his or her own evaluation of that data. Much will depend on who you think is trustworthy or gives a balanced account of the facts. It is a good exercise for people who are supposed, among other things, to practice the discernment of spirits.

Responsible believers concerned about nature need to recognize that, at the end of the twentieth century, religious emotion and idealism are far from sufficient for dealing with the environment. Many well-intentioned acts, not merely the previously mentioned Buddhist attempt to return captive animals to the wild, may themselves cause problems for nature or for human beings. Without scientific as well as theological knowledge, good intentions are blind. With only science, which has a history of separating human purposes and values from merely instrumental questions, there will be no recovery of a broad and deep connection with the fullness of the world as we experience it. Maintaining balanced judgment and religious wisdom about how these two human faculties mesh in specific circumstances will be the central challenge to religious people concerned about the environment in future years.

The Two Paths

Broadly speaking, there are two main approaches to environmental issues, whether they are religiously motivated or not. As Fred L. Smith, Jr., a former senior policy analyst at the Environmental Protection Agency, has put it with a touch of humor, the people who adopt one or another of these views might be characterized as the Terrible Too's and the Cornucopians. The Terrible Too's believe: "There are *too* many of us! We consume *too* much! We rely *too* heavily on technology, which we understand *too* poorly!"[1] Terrible Toos often predict an apocalypse unless we repent of environmental sins and take serious steps to mend our ways. Vice President Al Gore gave classic expression to this view: "Unless we quickly and profoundly change the course of our civilization, we face an immediate and grave danger of destroying the worldwide ecological system that sustains life as we know it."[2]

By contrast, the Cornucopians believe that human innovation, both in our technologies and in our institutional arrangements, will be more or less adequate to the challenges we face. They point to the stunning successes in boosting food supply through the Green Revolution, in water conservation, in reduction of harmful substances being released into the air and water, and in slowing rates of population growth as proof that human beings, endowed with intellect and will by God or, if you will, by evolution, can change their behavior, thereby avoiding the dire fate predicted by the Terrible Too's. For Cornucopians, overly strict restraint on technological advances, whether by national or international legislation, would freeze us in our current state and prevent the development of a more efficient, friendlier relationship between the human race and the world of nature. Rising levels of material well-being not only offer a better life, in this view, but also lead to slower rates of population growth and give societies the capacity to devote considerable resources to cleaning up past messes and to promoting environment-friendly practices.

There is, of course, no need to choose absolutely between these two approaches. It is not uncommon to see Cornucopians arguing for developments that will have a lighter impact on the earth. And the Terrible Too's — Al Gore is a good example — often become new technology junkies in quest of an environmental Grail. But religious people will want to situate all such approaches within larger perspectives. Some of the solutions suggested by the Terrible Too's, such as reducing activity and simplifying our ends or using different means such as public transportation and solar energy, may, in the right circumstances, be choices that religious people will want to make. The long spiritual traditions of asceticism could be joined with concerns about nature in a fruitful way if the asceticism is undertaken for the right reasons. Programs to educate children from their earliest years, if well supported by good science and a sincere desire to eliminate wasteful behavior (big "ifs" at present), could be an important dimension of the overall response to environmental challenges.[3] The Cornucopian perspective, too, might be inserted into the Western tradition of religious activism, not merely in the political sense that term has acquired in recent decades, but in the full biblical context of human work as willed by God. The entrepreneur who properly values nature may then become another kind of environmentalist—and a valued one for the new methods he or she will promote around the world.

The brief treatment of a few environmental issues below focuses largely on the United States and developed countries for several reasons. First, those of us who enjoy the fruits of prosperity have a special obligation to use our resources to lead the environmental movement. We possess not only material means but also crucial knowledge that must be applied to varying situations around the world. And as a first approximation of what needs to be done for people and habitats in straightened circumstances, a look at our own, often far easier situation may be helpful. This "hopeful interlude" cannot go into great detail about other parts of the world. But it is worth remarking that, in principle, less developed countries present no unprecedented problems. In practice, it may be more difficult to reform government and economic systems in developing countries than to identify practical solutions to local problems.

Higher consumption levels and population growth will continue, justly so, in developing countries for the near future. So we are faced with finding ways to take care of both humans and nature for many years to come. The aggressive population policies, for example, promoted by international agencies, governments, and private groups not only present serious ethical questions about who shall tell whom how many children to have, but also wrongly focus attention on what may be an increasingly marginal dimension of the problem. Fertility rates have been slowing around the world with few exceptions. Developed countries already have fertility rates below replacement levels; some parts of Europe show a demographic collapse, with an average of less than one child per family.[4] In the United States, population continues to grow slowly because of immigration and higher birthrates among immigrants.

But even developing countries have cut fertility rates almost in half in the past few decades from six births per woman to about three.[5] Everything in recent history leads us to believe that as material wealth rises around the globe, population will stabilize. Much, of course, will depend on individual and group decisions, but most experts agree that population is likely to stabilize at somewhere between eight and ten billion people in the coming century. As large as these figures seem — almost twice the current world population — there is good reason to believe that environmentally sustainable technologies exist in agricultural and manufacturing even today for feeding, housing, clothing, and educating that many human beings.

Bread for the World

Take, for example, the crucial question of food. Grain production will have to rise, as it has in recent decades, to keep pace with the population. But it would be wrong to think that feeding twice as many people will require twice as much farmland, with the accompanying destruction of wildlife habitat and ecosystems. According to agricultural scientist Paul E. Waggoner, a combination of techniques may actually make it possible in coming years to *reduce* the amount of land dedicated to cultivation, returning more of the earth to other uses or simple protected areas, even if the world population grows to about ten billion before stabilizing: "If during the next sixty to seventy years the world farmer reaches the average yield of today's U.S. corn grower, the ten billion [people] will need only half of today's cropland while they eat today's American calories."[6] Such a result will require a creative combination of new plant strains, planting and irrigation methods, pesticides, and fertilizers. Waggoner takes up each of these in turn and finds that, far from the usual image of modern farming as just another term for soil erosion, drought, and chemical poisoning, the new agriculture will reduce the negative effects of old techniques.

During this century, Europe and America have already reduced the amount of land they use for agriculture (and as a result, have been forced to subsidize farmers to keep land idle, since food is so abundantly available that price mechanisms alone no longer support many farmers). Globally, new methods have kept land dedicated to agriculture at the same level as at the middle of the century despite rising population. If similar methods are adopted elsewhere, as seems likely, Waggoner predicts that in sixty to seventy years, a reduction in the amount of land devoted to farming equal to the area of India will occur, even if population doubles!

Of course, experts disagree about the prospects, depending on whether they belong to the Terrible Too's or Cornucopian school of environmental thought. For instance, Lester R. Brown of the Worldwatch Institute has been engaged in a kind of running battle with his opposite numbers over various environmental questions including food production. The Worldwatch Institute has periodically issued a book-length report, *The State of the World,* to warn of current trends and to propose changes in population and human activity to offset environmental stresses on nature.[7] In recent years, that approach has been challenged by a publication in a similar format but from a Cornucopian perspective, *The True State of the Planet,* edited by Ronald Bailey. Comparing the data and analyses in these two volumes would be a useful exercise for anyone wishing to arrive at an informed position about the environment.

What does not differ in the two approaches, however, is the basic acceptance of intensive farming and other technological advances to deal with current problems. In another volume, *Saving the Planet,* Lester R. Brown and others from the Worldwatch Institute advance arguments quite similar to Paul Waggoner's and the Cornucopians' about the use of intensive farming to keep pace with population growth.[8] In this debate, even the Terrible Too's seem to have accepted that there are ways—sustainable ways as well—to meet both human and ecological needs. Brown takes occasional fluctuations in growth of grain production as warning signs and perhaps reads some data with unnecessary pessimism. For example, he regards as a sign of regress the fact that less land is devoted to grain cultivation, when it may actually be a sign of less acreage needed owing to better farming methods (as in the United States). He also tends to neglect political factors: the recent demise of the Soviet Union, a large grain producer, affected grain markets. But the price of food has been steadily declining almost everywhere that political turmoil has not interrupted agricultural advance. Famine is rare. The conclusion is plain: we already have the means to meet population demand for food, although distribution may remain a problem.

The Environmental Good News

Similar predictions on other environmental issues come from highly credible sources far less engaged in policy disputes than the Terrible Too's and Cornucopians. In the summer of 1996, a remarkable issue of *Daedalus,* the journal of the American Academy of Arts and Sciences, appeared. Under the overall title "The Liberation of the Environment," it gathered together several essays arguing that the most apocalyptic claims about damage to the earth's systems had been greatly exaggerated. In the preface, editor Stephen R. Graybeard characterizes the once formidable and environmentally alarmist Club of Rome as "an institution virtually ignored today." He suggests that early advocates of the "no-growth society" and the "stationary society" were promoting "the fanciful inventions of a group of utopians." Graybeard even repudiates a 1973 issue of *Daedalus* itself that had argued in favor of no-growth theories. Looking back on what we have learned in the past twenty-five years, he describes the pessimism of his predecessors as saying "something about the spirit of that seemingly now-remote age." In short, both the editor and the contributors to this prestigious journal regard the Cassandra-like warnings of the past two and a half decades to have been badly mistaken.[9]

Instead, they propose, on the whole, a much more optimistic and manageable view of problems and solutions:

> Environmental problems, as presently defined — with respect to energy and food, but in most other instances as well—derive from the very specific scientific and engineering feats of recent decades. Their resolution depends on the same human intelligence, to a very considerable degree.
>
> Pollution is controlled today in ways that would have been inconceivable a few decades ago. So, also, in both agriculture and industry, there is a concern with productivity and efficiency, with avoiding excesses that were once thought wholly legitimate, almost unavoidable, given the state of technology at that date. Manufacturing, like agriculture, is a transformed enterprise, with materials, wastes, techniques, and hazards perceived in ways that were not at all common when either Theodore Roosevelt or Franklin Roosevelt sat in the White House.[10]

Such confidence about the environment coming from a respectable and not at all unconcerned source may herald a much less inflamed and far more realistic attitude towards the environmental issues we face on the eve of a new century.

There may be unpleasant surprises down the road — there are no guarantees about the future. But as we know from recent history, there will be some good surprises, entirely unanticipated, as well. No scientist or engineer predicted the major discoveries of this century that have changed human life. We may expect similar surprises in the centuries to come. The central factor that seems to unite all the *Daedalus* contributors is their belief that, on balance, technology and civilization will in the future lead to a *liberation* of the environment. More specifically, as Jesse H. Ausubel of Rockefeller University puts it, "the liberator of our title is human culture."[11] Human culture promises in the future to move us from using carbon-based energy sources to non-polluting hydrogen systems. Our culture has already enabled us to produce more for less, and developing nations could benefit from advanced technologies without having to go through the environmentally unsound earlier phases of industrialization (though how to carry out this technology transfer remains a vexed political and economic question). In fact, much of modern cultural transformation involves a dematerialization of activity by means of computers and other devices that would have been unimaginable just a short time ago.

Ausubel points out that all this also constitutes a liberation for human beings. More people do die from environmentally connected cancers today than in the past, if we take "environmentally caused" to mean induced by toxic chemicals, radiation, or other human factors. In America, about 5 per cent of cancers are so caused.[12] But this must be viewed against a background of deaths that resulted earlier that were "naturally" environmentally caused. Ausubel cites typhoid, cholera, tuberculosis, diphtheria, influenza, pneumonia, measles, and whooping cough as just some of the forms by which natural environmental threats killed us not long ago. Cancer may have a long development time, and rates from exposure to chemicals or radiation a few decades ago may still go up. But at the same time, it is clear that such exposure peaked long ago and will increasingly be a thing of the past.

Once dreaded threats from diseases and wild animals have become relatively weak with human advance, so that we now have the luxury of imagining wild areas like forests as benign conveyors of spiritual uplift. Societies in the past, with a much more concrete experience of undisturbed nature, would have found such views unrealistic: "Jungles and forests, commonly domains of danger and depravity in popular children's stories until a decade or two ago, are now friendly and romantic. . . . The characterization of animals, from wolves to whales, has changed. Neither the brothers Grimm nor Jack London could publish today without an uproar about the inhumanity of their ideas toward nature."[13] There is, of course, nothing wrong with this cultural shift, just as long as we recognize that it is the very success of human culture in warding off natural threats that now enables us to afford more generous impulses toward parts of nature that were once deadly foes.

One of the unavoidable paradoxes of human existence is that every solution gives rise to further problems. The environmental issues of the next century are unlikely to resemble the ones we face today. Not that long ago, for example, many environmentalists worried about how we would dispose of mushrooming quantities of trash and sewage. Those problems no longer exist thanks to innovations in handling such wastes. But let us briefly examine a few of today's more prominent environmental issues in an attempt to see how they may appear tomorrow.

Acid Rain

In the late 1970s, one of the greatest environmental scares in Europe, Canada, and the United States was the prospect that forests and lakes would be destroyed by "acid rain." Environmental writers predicted the demise of vast tracts of woods in developed countries, and some news organizations were even persuaded to show areas that they believed revealed an ominous development. To oversimplify the science somewhat for the sake of brevity, as the problem was first understood, it was assumed that sulfur dioxide and nitrous oxides emitted by industrial plants, especially coal-burning power plants, were turning into sulfuric and nitric acid in the atmosphere and then chemically "burning" pine needles and leaves on trees, killing off large tracts of forest. Those same acids, it was thought, were also burning crops, potentially harming human lungs, and making lakes highly acidic and therefore incapable of supporting fish populations. The situation seemed so dire that in 1980 the U.S. Congress voted to spend an astonishing half a billion dollars over ten years for the National Acid Precipitation Assessment Program (NAPAP).

NAPAP reported, however, that there was no reason to believe that acid rain was a widespread and serious threat to trees, human beings, crops, or lakes. Indeed, some compounds were benign; nitric acid, for example, occurs naturally and acts as a fertilizer for plants. Damage from human-generated acids was slight. Something like 3 per cent of Eastern forests—the trees believed to be most at risk because they are downwind of Midwestern power plants that burn high-sulfur coal — showed some damage. And only 4 per cent of lakes had high acidity, some perhaps because of reversion to previous conditions before agriculture had temporarily made them more alkaline. (Florida contains a large number of acidic lakes for reasons having nothing, it seems, to do with acid rain.) Some sections of the southern Appalachians were hard hit because of a unique combination of pollution patterns, altitude, precipitation, and the presence of vulnerable species such as the red spruce. But NAPAP arrived at surprisingly optimistic conclusions.

This does not mean that there is no acid-rain problem. It was a good thing that the United States and other countries began in the 1970s to reduce emissions of sulfur and other contributors to acid rain. Today, these emissions are about 95 per cent less than they were at the start of the programs. That step was costly, probably far more costly than the problem it was supposed to remedy warranted. But given the synergies in nature and our technological ability to stop needless pollution, it may someday pay dividends. All this was accomplished despite various vested interests: miners, utility companies, factories, and politicians. Forests have been expanding in the United States in recent decades, as the amount of land devoted to agriculture decreases and wilderness areas are placed in reserve by private or governmental institutions. Clean-air provisions will, in all likelihood, mean that forests generally will be healthy for a long time to come.

Gregg Easterbrook, who is an environmentalist but a rather optimistic one by current standards, has remarked that the debate about acid rain is emblematic of a wide range of environmental problems:

> [Acid rain] is a genuine environmental problem, but its significance has never been as important as claimed. Meanwhile the cures, some already being put into effect, are effective and for the most part affordable. . . . The above description can be applied to nearly all current environmental problems: genuine but exaggerated, subject to correction surprisingly quickly at reasonable cost. Thus understanding acid rain provides a model for understanding the reality, as opposed to the hype, of most environmental issues facing humankind.[14]

How has it happened that we find ourselves in this favorable position? No small credit is due to the environmental movement, though to read the literature of environmental groups, you would rarely get the impression that there have been any successes.

Up a Tree?

If the acid-rain problem and its solution were relatively simple, many other environmental questions are not. Discrete parts of nature —whether rain forests, oceans, streams, or wetlands—present complexities that outrun our present capacities for scientific study. Yet even for the more complex cases, the situation is not beyond roughly prudent management. For example, the harvesting of trees for woods and paper has often been presented as threatening irreparable damaged in developed countries. Irreparable damage is rare in nature. America's northern forests grew quickly from northern lands scraped bare by the glaciers of the last Ice Age. Fire and other natural causes regularly clear large areas. "Old growth" forests, which sound like almost mythically ancient habitats, are typically only a century or two old, and may regenerate (as is happening in many parts of the country) if stands of trees are left alone for a while.

At present, the forests managed by private companies are more of a solution to environmental worries than a problem. Almost no company today practices the aggressive clear-cutting of the past. In fact, both economics and ecological concerns make intelligent planting and harvesting of trees in what amount to large lumber farms a lighter influence on nature. Young trees absorb more carbon dioxide from the air than older trees (a side benefit that may help compensate for industrial release of carbon dioxide into the air). Young forests typically provide varied habitats for larger numbers of species than the older, more uniform forests contain. Despite some fragmenting of habitats, forests in the developed world now are in good shape and can be sustainably harvested and protected for a long time.

That is not the case in other parts of the world, however, and worries about trees remain valid there. In addition, tropical forests and rain forests contain large numbers of exotic species that are threatened by poor forestry or aggressive use of slash-and-burn agriculture. And lurking behind all such problems are the poverty and political turmoil in many developing nations that make application of demonstrably sustainable methods problematic.

Yet even here, progress is being made. Until the late 1980s, the Brazilian government, which oversees the Amazon, one of the largest and most important rain forests in the world, paid settlers to clear land or made land clearance a condition of ownership. When Brazil was trying to settle its vast interior, that policy may have made sense. But in the last decade, protecting the forest from environmental degradation has taken precedence in Brazilian policy over the minimal gains from agriculture. Environmental literature often connects that agriculture with raising cattle for American companies like McDonald's. Even if true, the connection is a weak one and likely to disappear. There are better ways to raise beef without touching the rain forests. The question remains, however, what economic alternative to the destruction of rain forest can be offered to poor nations with few other options.

Yet even at the time when the Amazon was being furiously attacked, nothing like the vast clearings predicted by environmental groups ever occurred. Some predicted that half of the Amazon would be burned down by 2000. In fact, 90 per cent of the Amazon is still untouched, only a fraction of a per cent is lost to agriculture yearly, and methods are being changed to improve the situation all the time. In 1989, the Brazilian government stopped its subsidies to forest burners. If this can occur in relatively poor Brazil with its long history of special-interest politics and corruption, it can occur elsewhere if governments will take the initiative. In areas like Southeast Asia, where rain forests are being damaged by reckless harvesting of prized hardwoods for sale to Japan at the very moment when economic downturn is putting pressure on the region, some special arrangements may be required.

Endangered Species

As we have seen in earlier chapters, nature quite casually brings species into existence and then drives them extinct. The very forces that open up new niches for species may later close them off. Human activity threatens some species whose natural time has not yet come, but nature itself has already eliminated all but 1 per cent of the species that have ever existed. Remarkably, some scientists believe that since the end of the last Ice Age, the earth has blossomed into the greatest biodiversity in its history. Despite natural extinctions, we are, then, the heirs of an astonishing natural wealth of biological species and need to be aware of our impact on creatures we may not yet even have identified.

Data about the number of extinctions occurring are notoriously unreliable. The problem begins with the number of species that we think exist. Today, we know that an intricate interaction of various species produces the food, air, water, and materials that we need. In the great diversity of life around us, we also find forms that move our hearts and imaginations. Human beings, no doubt, first dreamed about flying by watching birds, and a host of other things we may think the human mind simply invented stem from our observation of nature. Species range from simple one-celled organisms up to large and complex creatures such as whales. Yet until 1677, when van Leeuwenhoek invented a simple microscope, we did not even know that single-celled organisms existed. Discoveries about the importance of nitrogen-fixing bacteria in fertilizing soil and of carbon-dioxide-consuming plankton and other ocean life came later. As our knowledge of the earth has expanded, so has our estimate of the total number of species. Edward O. Wilson, a biodiversity specialist, believes that a single tree in a tropical rain forest may contain as many species as are known to exist in all of England.

Wilson and a few other scientists estimate that for every naturally occurring extinction there are perhaps thousands of human-related ones. Yet globally such large-scale die-offs have been difficult to confirm. They would probably be quite well documented in this country. Elsewhere, especially in rain-forest conditions, they are largely a matter of speculation. Wilson and another scientist, Norman Myers, have been the most prominent doomsayers. If they are right, as seems likely, that rain forests contain thousands of species per acre, even the smaller percentage of rain-forest loss indicated by recent studies means tens of thousands of species will disappear without our certain knowledge of their existence. Two non-scientists—Julian Simon and Aaron Wildavsky—have made a strong case that Wilson's and Myers's methodology and predictions are without solid scientific foundation. Myers and Simon jointly wrote a book, *Scarcity or Abundance?*, in which they debate a variety of environmental questions including species extinctions.[15]

Whatever may be the case in the rain forests, Wilson and Myers seem wildly inaccurate in their speculations about better-known areas. Gregg Easterbrook has observed:

Taking into account Wilson's estimate of 137 species extinctions per day and the land area of the Northwest forest belt, I roughly calculate that in the region in question 75,000 species should have fallen extinct during the postwar era. Yet no extinctions have been observed. Certainly some uncataloged species might have slipped to extinction unnoticed. But 75,000 overlooked extinctions in an intensely studied region? Not bloody likely.[16]

We have, instead, much better indications that the nature we once harmed is capable of rebounding. For instance, about seventy species were put on the endangered-species list when it was first compiled, in 1973. To date, twenty-five years later, it appears that seven species have indeed disappeared. That is 10 per cent of the most vulnerable group, which may appear a lot or a little depending on how you read the numbers. For most of us, any number of needless human-caused extinctions is too high, despite the fact that extinctions occur regularly in nature. But the decades before 1973 were probably the years in which the human race put the greatest pressure on species and habitats that they are ever likely to experience. And the documented cases of species disappearance were few.

It is useful to recognize the relative importance of various factors in producing species extinction. Besides natural causes, the worst factor in documented extinctions is the intentional or unintentional introduction of non-native species into a habitat, which has caused 39 per cent of known extinctions. Habitat destruction comes next with 36 per cent, and hunting has extinguished 23 per cent. Only 2 per cent of extinctions have occurred because of pollutants, pesticides, or other causes.[17]

Steps have already been taken to ban or restrict the use of certain pesticides, to protect habitats or prevent their fragmentation, to regulate fishing and hunting, and even to intervene when nature itself seems to be leading species towards extinction. Pressure from people, of course, continues, especially in the developing world. But scientists believe that something like 75 per cent of the land on every continent except Europe could be reserved for wildlife. Long-cultivated Europe itself is not particularly a threat to species, a sign that civilization and nature may complement each other. Today, eagles, falcons, several kinds of wolves and whales, and a variety of other species are in better shape than when we first took notice of their plight. Even elephants and rhinoceroses are flourishing in some places where creative management of wildlife has been applied. The battle over protecting species has not ended and there have been significant casualties along the way, but it would be difficult to say that the future is likely to be harder on nature than the past.

In early 1997, a report in the journal *Science* claimed that endangered species tend to be clustered around a few sensitive places in the United States and other countries.[18] Less than 2 per cent of our nation's land provides the habitat for over half the endangered plants, molluscs, insects, fish, birds, and mammals. Attempts to preserve biodiversity, the scientists said, could concentrate on these areas with great effect. In the United States, Florida, California, and Hawaii show the greatest threats to species.

Extinction often looms for what scientists call "endemic" species that exist in only a small range. Human intrusions into these areas, especially urban development, amount to "the formula for extinction these days," according to Dr. David S. Wilcove of the Environmental Defense Fund. Naturally, any serious desire to preserve species would have to take into account more than habitat issues; at the very least, it would have to address the causes of endangerment in the first place. Dr. Dennis D. Murphy, president of the Society for Conservation Biology, described these findings as evidence that, contrary to some misconceptions, "the entire country isn't overburdened with an imperiled species challenge."[19] Despite other worries about biodiversity, this is certainly good news.

The New Agriculture

As we mentioned earlier, one of the primary reasons why forests are returning in many places is that less land is needed for agriculture today. Here is where we see the kinds of trade-offs that must sometimes be accepted in our relationship with the environment. Given the large growth in human population in this century, we might have expected a doubling of land devoted to growing food. That did not happen primarily because of two factors: first, increased yields from new strains of plants and intensive farming techniques, and second, the success of fertilizers and pesticides in boosting and protecting foodstuffs.

Chemical fertilizers and pesticides have gotten a bad name because in earlier decades they were used indiscriminately and without proper testing of their effects on human beings and wildlife. At present, it appears that there is not, and probably never has been, any major threat to human beings from either group of chemicals, properly used. Pesticides did temporarily depress the reproductive rates of some fish and bird species, particularly animals like eagles, falcons, and songbirds. After Rachel Carson sounded the alarm about the effects of DDT in particular, its use was banned. DDT is a persistent toxin that unfortunately remains in small amounts in some places where it was earlier used, but its harm to wildlife today is minimal, and the advantage to wildlife of less dangerous pesticides is incalculable.

A single pesticide or preservative may prevent a third of a given crop from being eaten by pests or spoiling in transit, making it possible to keep an equivalent amount of land in a relatively undisturbed form for natural species. Furthermore, some new techniques such as integrated pest management can reduce or, in some contexts, eliminate the need for pesticides by using natural predators like ladybugs and praying mantises. Natural pesticides present in certain plants are being adapted to deal with bugs in another way. Even crop rotation is back as a way to reduce the number of pests in off years, with corresponding large yields when the crop returns. And farmers have learned in most cases how to make sure that pesticides and fertilizers do not run off into streams or other places where they may do harm.

It used to be thought that the use of pesticides would lead to strains of superbugs resistant to all pest control. To a large extent, that fear has not been realized. Yet even if it were to come to pass, it would be the kind of change that nature itself constantly produces, with new species seeking environmental niches until counteracting forces arise to check or eliminate them.

Toxic Wastes

Similar to the pesticide and chemical-fertilizer scares were the alarms raised in the 1970s and 1980s about toxic chemicals from industrial processes. Environmentalists rightly claimed that for decades factories had simply been dumping their wastes, often without concern about where those chemicals would wind up and what effects they would produce. As with many other environmental issues, the criticism led to rapid changes in policies. The most worrisome substances released into the environment by human industry included heavy metals, dioxin, PCBs, and some chlorine compounds. Radioactive wastes and mine tailings formed another large group.

The best news is that all these substances are now regulated quite carefully along with hundreds of other chemicals, even though many such chemicals pose no known danger to humans. In the merely good category are findings that, with only a small number of exceptions, the chemicals we foolishly allowed into nature in earlier decades have had relatively small harmful effects on humans. These findings are in remarkable contrast to the scares we have been exposed to in recent years over substances such as Alar and dioxin. Each of these has the potential to cause harm, and it was good that the highly exaggerated claims about them got us to look more carefully at what we were doing. We have already seen how restriction of DDT use has largely removed it as a concern. But it is worth looking at another substance, dioxin, for what its history in the environmental debate says about our reactions.

Dioxin burst into national consciousness when the mother of a young boy blamed his epilepsy on the fact that the land on which their house sat at Love Canal, near Niagara Falls, New York, had earlier been contaminated with dioxin. Regrettably, the only reason that she and her neighbors were in the chemical zone was that local officials had earlier ignored warnings from the chemical company that once owned the land that potentially hazardous wastes were buried there. The town allowed houses and a school to be built. When the chemical fears about contamination arose, officials, running to the opposite extreme, evacuated the area and started a costly relocation and cleanup. Yet much of this may have been unnecessary.

As bad as exposure to dioxin was and is, fears that it would have harmful human effects were based on flawed animal studies. Those studies were also used to support the claims of some Vietnam veterans that exposure to Agent Orange, a dioxin-containing defoliant widely used during the war, was responsible for various health ailments. Evidence for the veterans' claims was not clear. Some had documentable health problems that were hard to correlate scientifically with exposure to dioxin. For good political and moral reasons, they were granted benefits. With Love Canal, it appeared that people living in a normal, middle-class American neighborhood could have been inadvertently exposed to dangerous, wartime chemicals.

But after the initial scare, cooler heads prevailed. It soon became clear through further study that, though still not a substance to be taken lightly, dioxin did not have the monstrous effects thought. In fact, when a chemical factory in Seveso, Italy, exploded in 1976, it exposed residents to perhaps fifty times the amount of dioxin that the average Vietnam vet had received. Naturally, scientists and doctors have followed this unfortunate group quite closely. Fifteen years after the accident, epidemiological studies showed a tiny increase in some problems, but basically the health of the people in Seveso was virtually indistinguishable from that of others in the region. Some cancers can take a long time to develop, and dioxin and other compounds can have cumulative and persistent effects. But so far, the record is mercifully benign, and elimination of dioxin exposure means diminishing ill effects in the future.

Many other substances fall into similar categories. It takes exposure at certain doses for chemicals to harm life, including human life. For the most part, industrial toxins are in restricted areas where few people are exposed to them, and those distributed more widely are diluted enough that they pose little risk. The U.S. government has been trying to clean up the more serious industrial and military sites through Superfund, sometimes at costs far higher than any possible benefit. If millions and even billions of dollars are spent dealing with toxic-waste problems that most scientists now regard as second- or even third-tier priorities, less money is available for research that may have a much greater impact. For example, it has been estimated that to save one life from the ill effects of asbestos costs over $110 million. By contrast, the United States spends an average of only $800 on research per person who gets breast or prostate cancer. Given that environmentally caused deaths are demonstrably few and deaths from breast and prostate cancer many, the disproportion between the figures reflects a political rather than a scientific judgment about risks.

There are some persistent toxic problems mostly having to do with sites that cannot be easily restored with current technologies. Some substances, like nuclear waste and mine tailings, can only be contained, not adequately neutralized, by current technologies. But on the whole the current situation is clear. We have passed the point where we thought that the world was large and our chemical impact on it negligible. We have taken steps to prevent the worst forms of toxic emissions and are dealing with the effects of past mistakes. We are not in a bad position at the moment. In the future, we will be in a much better one.

Ozone Depletion

Ozone plays a variety of roles in the biosphere. At high altitudes, it filters out ultraviolet rays from the sun that can cause sunburn and skin cancer. At low altitudes, especially when combined with auto exhaust and high temperatures, it produces smog. We have made enormous strides against the problem of low-altitude ozone. Until the extremely high temperatures that were the result of El Niño in the summer of 1998, all of the formerly smog-afflicted cities in America — even Los Angeles — were showing steep and regular declines in smog and many other forms of air pollution, at a time when population, the number of cars, and local industries were growing at a fast pace. Smog will likely soon be a thing of the past, though other problems such as higher asthma rates in cities present some new questions about air pollution.

The high-altitude ozone problem, however, appears likely to remain a concern, if a manageable one, for the near future. The theory behind the observed thinning of the ozone layer is that chlorofluorocarbons (CFCs) released into the atmosphere gradually work their way into the stratosphere, where they break down, allowing free chlorine to attack the three oxygen atoms that form ozone. Unfortunately, one chlorine atom may chew up thousands of ozone molecules, and the effects persists for a long time, since CFCs are highly stable and do not easily wash out of the atmosphere with rain or other natural processes. Around the poles the problem has been particularly troublesome because a combination of conditions, including especially cold temperatures at certain seasons and, in the south, the Antarctic vortex in the high atmosphere, exacerbates the effects.

Yet part of the problem has been solved. In the 1990s, a worldwide ban on the use of CFCs for purposes such as aerosol propellants and plastic-foam production went into effect. The remaining—and large—problem is what to do about the CFCs that are used in refrigeration units: car and building air conditioners and the cooling mechanisms in refrigerators and food-transport vehicles. Substitutes are available for the CFCs, but they present other problems such as flammability and are quite expensive. For example, it is estimated that it would cost $1,000 to retrofit a car to use the alternative coolants. Multiplied by millions of cars and other refrigerating units, the cost would be staggering.

Yet there are reasons to doubt whether the crisis, as usually described, is occurring. Careful measurements of ultraviolet radiation (UV) reaching the earth's surface by the National Oceanographic and Atmospheric Administration and other credible bodies have shown no increase but a slight *decline* in the last decade. And some of the alarms about "holes" similar to the Antarctic one, which existed long before human causes could have produced it, have proven unfounded. Recent measurements of ozone depletion may have more to do with natural fluctuations and sun cycles than previously believed.

Furthermore, even the worst-case scenarios overstated the threat. By one estimate, the amount of ultraviolet radiation that would have resulted from a 15 to 20 per cent loss of stratospheric ozone would have been the equivalent of a person in the Northern Hemisphere moving sixty miles further south, since the strength of the sun and the amount of UV radiation that gets through the atmosphere vary with latitude. Unfortunately, stories began to appear in the media, and were repeated by Al Gore in *Earth in the Balance,* that some animals and fish at the poles and in Argentina were going blind because of the higher radiation.[20] This proved wildly untrue. Animals are adapted to fluctuations in ultraviolet radiation. And where documented, the blindness Gore reported seems to have been the result of conjunctivitis, not anything to do with sunlight. Similar horror stories about the effects of ultraviolet radiation on plants and humans also proved unfounded. Living things on earth have survived much higher UV levels in the past and have probably acquired resistance to the levels of change feared.

But as with the other issues examined here, we see a pattern of recognition of a problem and then attempts — sometimes exaggerated, sometimes appropriate — to deal with it. Ozone depletion does not appear likely to be a serious problem in the coming century. Yet it is useful to have learned the lesson that the complexity of the biosphere should make us vigilant about the immediate and remote effects we may have upon it in unexpected ways.

Global Warming

Global warming comes last in this survey because for many people it remains an unknown with apocalyptic possibilities. Indeed, global warming has probably become the premier environmental question because the other disasters we anticipated now seem unlikely to occur. In addition, global warming has unusual characteristics. If its causes are anthropogenic, they are not, strictly speaking, pollution. Carbon dioxide is a natural compound in the atmosphere. Oxygen-breathing animals give it off as a by-product; plants take it in and, through photosynthesis, provide themselves with energy and food and recycle oxygen back to those of us who breathe it. Methane, another natural greenhouse gas, is given off in large quantities by rice paddies and wetlands as well as other natural sources. The problem here is not pollution, then, but an imbalance of naturally existing substances. All credible scientific sources show that levels of atmospheric CO_2 have increased sharply since the industrial revolution, from about 220 parts per million (ppm) to 350 ppm; methane has risen from 650ppm to 1700 ppm. In the same period, the earth seems to have warmed about one degree Centigrade.

Everyone agrees about these facts. But there is probably no environmental issue more hotly contested than global warming, beginning with whether it is actually occurring. In the popular media, global warming is presented as being confirmed by scientists, celebrities, and even mere common sense. After all, haven't the summers lately been unusually hot and the weather strange? Anecdotal evidence, of course, is unreliable, especially when people have been led to believe that they are seeing something. But weather and climate are two different things: it may rain for a week in Seattle or Miami, but overall the two cities have very different climates. Among scientists, the question of global warming is much more disputed than may appear in media reports.

The U.N. Intergovernmental Panel on Climate Change (IPCC) issued a report in 1996 after serious study, and its weak conclusion was that "the balance of the evidence suggests a discernible human influence on global climate." Even this depends on computer-generated general climate models (GCMs) that the scientists themselves will tell you are quite speculative. Perhaps the biggest problem with these GCMs is that they do not predict what we know, let alone what we do not. When the models are set back to conditions a hundred years ago and the documented variables are added, the run forward does not come close to the climate we observe today.

And where do we find a baseline against which to measure average global temperature? As we saw in previous chapters, the earth is constantly changing and has experienced stretches of hot and cold centuries within the last thousand years. The end of the last century was both the end of a "little ice age" and, anomalously, a relatively cool period. Measurements are usually taken using 1880 as a baseline, when records were first systematically kept. From the 1940s to the 1970s — well into the age of human-generated carbon dioxide — we passed through relatively cool decades. The fact that 1977 was the coldest single year in this century may have had something to do with the fear of global cooling that was common at the time. The 1980s started an upward trend, and it was warm through the 1990s, one of the reasons for recent fears of global warming.

But even these data are uncertain. Ground-based stations, often operating within the heat dome generated by urban population centers, have been showing slight increases in average temperatures. Satellite measurements show a slight cooling. Indeed, in the summer of 1998 when El Niño was bringing devastating heat to crops in the United States, NASA's web site claimed that global temperatures have been declining by 0.06 degrees Celsius per decade.[21] A majority of scientists are quite skeptical about the scientific evidence of global warming. In a now famous 1993 poll of the members of the American Geophysical Union and the American Meteorological Society, the Gallup organization found that 49 per cent of these climate professionals believe that no anthropogenic warming has taken place. One-third said that the data were inconclusive. Only 18 per cent thought anthropogenic warming had occurred. The prestigious journal *Science* has repeatedly carried stories skeptical of the common view. The usual claim of global-warming advocates is that any scientist who questions the alleged facts must be in the pay of oil companies or others who benefit from fossil-fuel use. The truth, it appears, is that this is just not the case.

Nonetheless, caution is warranted. Climate involves too many factors to make predictions certain either for or against warming worries. We don't know how much other systems like the oceans, forests, and plants can adjust to absorb additional carbon dioxide, nor do we know how much atmospheric circulation patterns, or variations in the sun's intensity, or a host of other factors influence our situation. Such warming as has existed to date seems to occur mostly in the winter and at night, when any effects are likely to be benign (there may be an accidental self-regulating effect that joins human societies and the natural world in this development if it leads to lower fuel consumption for heating during winter months). Water vapor is the most important greenhouse gas of all, accounting for 99 per cent of the greenhouse effect, and there is no predicting the effect that current temperatures will have on evaporation. Some scientists speculate that more water vapor will trap more heat. Others theorize that increased cloud cover will reflect more heat back into space during the day and trap more at night, something that might be consistent with observed nighttime temperature increases. But at the end of all this speculation, we must say that the prudent course will be to find ways to reduce the use of fossil fuels and develop technologies that will be more efficient and ultimately eliminate our dependence on forms of energy that add to carbon dioxide in the atmosphere.

The Religious Perspective

As the above survey of current environmental questions shows, there is probably no environmental problem on the radar screen at present that is intractable. Nothing in this generally hopeful reading of our current situation, however, should lead anyone to think the struggle is over. We will have to continue strenuous efforts to produce more with less and improve environmental effects of human activity for a long time to come. And we know that we now have the potential to do great harm to ourselves and nature unless we are very careful. A great deal of the success we have had in handling environmental problems reflects the old human capacity to recognize and respond to new challenges. It takes an effort of imagination and creativity to live in the world we find. Some of our former errors were the product of sheer inattention. And that raises a question: does our vision of what the world is have any bearing on what we attend to? Clearly, pragmatic approaches are fine for many, perhaps even most, environmental problems. But the larger questions of where human beings come into the picture and how practical activities fit into the larger themes of meaning and hope take us to the heart of the religious approach to the environment. In the next few chapters, we will take a critical look at several of the principal approaches proposed for our situation as those have been presented by key religious figures.

3

Back and Forth:
The Cultural Contradictions
of Technology

Even in the brief outline we sketched in chapter 2, the surprising cosmology discovered by twentieth-century science suggests some fruitful, if tentative, paths for new thinking about our place in nature and our relationship with the environment. Yet few theologians have taken up the challenge of incorporating the new scientific truths into religious thought and imagination. To be sure, theology and science are separate disciplines, as we have seen, and there can be no question of simply using modern science to prove theological truths or of using theology to deny scientific truths. It was a recurrent cultural error over several centuries, on both sides of the divide, to believe that empirical science and speculative theology could somehow refute each other. The trauma for theology caused by the emergence and dominance of the old mechanistic Dynamo and the hubris of modern Western science and technology towards any theological claims have blocked any new vision of how the new non-mechanistic Dynamo and a renewed Virgin might, in respectful dialogue with one another, lead us into a different appreciation of the nature of our world.

This is unfortunate for several reasons. Not only are religion and science impoverished by their isolation from each other; but also each displays certain tendencies to extremism and instability as a result of an exclusive focus either on mere physical processes or on purely spiritual truths. As we shall see, the absence of a vibrant engagement with current science by more orthodox religious figures is leading to a proliferation of fantastic constructs by the less orthodox and the outright heretical.

Heresy has acquired a certain cachet in the last century or so, as if the departure from accumulated wisdom were a guarantee of daring and originality. Just as often, it winds up in confusion, self-contradiction, and subservience to current nostrums. *Heresy* derives from a Greek verb meaning "to pick out, to choose," from the full truth. Far from suggesting personal originality or the daring discovery of new truths, then, it means a narrowing of the world to the things we arbitrarily decide to select from the richness of reality. Given human nature, this almost always means constructing a personal fantasy that we prefer to the real world. For example, highly popular New Age currents seem to play on people's wishful thinking and gullibility in the face of the unknown. But there are similarly unfortunate projects under way in environmental discussions even by people who might be expected to be forewarned about the consequences of unorthodox conceptions of science and religion.

Father Thomas Berry: A New Revelation?

For instance, Father Thomas Berry, a Passionist priest, has exerted an enormous influence on religious thinking about the environment in recent decades. A cultural historian of impressive breadth and, at times, with a poet's gift of making the new cosmology come alive, he has elaborated one of the most powerful and well-informed revisionings of nature to be found in the environmental debate. His views have been adopted by several religious orders. And Al Gore, in line with his own apocalyptic fears, has cited Berry's authority for the contention that our civilization is rapidly destroying species, so that one-half million to a million species will disappear by 2000. Although no credible biologist has actually observed any such die-off (confirmed extinctions are somewhere in the hundreds for all of modern history, with a few thousand more species currently threatened), Berry predicts a catastrophe even worse than the Permian extinctions of 220 million years ago, when at least 90 per cent of species went extinct. Followed by Gore, he also proclaims that within our lifetimes we will bring about the end of the Cenozoic Era, which began 65 million years ago.[1]

Unlike many environmentalists, however, Berry believes that the old Newtonian vision — with its bias towards mechanism and its heedlessness of the effects of human action on nature, as those traits may be observed in Bacon and Galileo and were institutionalized in the capitalist, industrial, and democratic revolutions — has turned into an astonishing new revelation: "If our science has gone through its difficulties, it has cured itself out of its own resources. Science has given us a new revelatory experience. It is now giving us a new intimacy with the earth."[2] But this new story is resisted by an unprecedented cultural pathology "deeply imbedded in our cultural traditions, in our religious traditions, in our very language, in our entire value system."[3]

Particularly in his collaboration with the mathematical cosmologist Brian Swimme, Berry has succeeded better than any other recent writer in presenting the cosmic story as we now know it from science as a basis for a coherent myth that could inform the religions and cultures of the world. Their jointly written volume, *The Universe Story,* puts into accessible and sometimes evocative language how the "primordial flaring forth" may be understood to explain our origins, our ongoing relationship to all things, and our responsibilities toward creation.[4] Berry has elsewhere summarized our situation as calling for a reduced human presence: "The time has come to lower our voices, to cease imposing our mechanistic patterns on the biological processes of the earth, to resist the impulse to control, to command, to force, to oppress, and to begin quite humbly to follow the guidance of the larger community on which all life depends."[5]

Humility always exerts a strong religious appeal, as it should, given human tendencies to self-aggrandizement. But from the outset, certain persistent and troubling features crop up in Berry and Swimme's work. First, they began by dismissing virtually all existing religious traditions as inadequate to a task they see as apocalyptic in its implications: "The existing religious traditions are too distant from our new sense of the universe to be adequate to the task that is before us. We cannot do without the traditional religions, but they cannot presently do what needs to be done."[6] Instead, as in much environmentalist literature, native peoples are recommended as wise guides, though why they are more relevant than traditional Western religions to our unprecedented situation is not clear, to say the least. Perhaps in recognition that ecology has risen to an importance above all other concerns in his own religious thought, Berry has taken to describing himself, not as a theologian or priest, but as a "geologian."

A Wholesale Indictment

If all of this were merely a statement to the effect that we face something new that demands every bit of accumulated human wisdom and ingenuity to solve, the point would be an obvious one. Berry, however, has constructed a history of the planet that indicts virtually all previous human cultures, especially those developed out of the Bible, for their contributions to current environmental problems. Prehistoric societies, whose archeological traces are few and of disputed meaning, are the ideal; presumably they lived easy on the land and with one another, and experienced themselves in their everyday lives and rituals as part of the cosmic process. But every culture that has left a rich record comes in for no little indictment. Given that basic historical perspective, Berry naturally does not view environmental problems as amenable to piecemeal remedies; rather, they constitute a worldwide crisis that will call for the radical reversal of our attitudes and a restructuring of most of the political, cultural, economic, technological, scientific, and industrial institutions on which we have come to rely. For him, the problem is cosmic and the solution must be cosmic as well.

Berry's reading of our historical relationship to nature and development as a species is remarkably narrow for a man otherwise sensitive to the complexities and ambiguities of human existence. The gulf between his understanding of human nature and his program for nature is so large that it is difficult not to feel that something other than theological analysis may be at work. Like the present, our past was human in its weakness, error, and corruption. But it may also have embodied certain workable, if imperfect, compromises that even the most visionary of contemporary thinkers might have a hard time discovering all on his own. Human experience, for instance, reflects a constant struggle with nature. Yet Berry believes, contrary to such experience, that "the earth will solve its problems, and possibly our own, if we let the earth function in its own ways. We need only listen to what the earth is telling us."[7] Berry even calls for the emergence of lawyers who will possess "a sense of the inherent rights of natural beings."[8] As we have seen, the best new science of the biosphere gives little reason to think that natural processes are generally in harmony or that, left to themselves, they will lead to desirable human outcomes. And nature itself is certainly no respecter of the rights of natural beings or even whole species. Yet Berry's religious enthusiasm about the nonhuman universe appears to have narrowed his vision with regard to our empirically verifiable world. Like many in the environmental movement, he views human beings as a flaw in what he believes to be an otherwise admirable realm: "We are the affliction of the world. We are the violation of the earth's most sacred aspects."[9]

Moderate and traditional theologians who work on environmental issues often try to characterize figures like Berry as extremists and to argue that taking his views as somehow characteristic of the movement as a whole gives a false impression of what most religious people think about nature and man. It is true that the kind of radical rejection of almost all known human societies in Berry's work represents a minority position among religious environmentalists, but only in the purity of its radicalism and the degree to which it would blithely cast off current social arrangements for untried experiments. Berry's basic vision of a gloriously unfolding cosmos, where the earth is primary and man is derivative and of uncertain intrinsic value,[10] has become the implicit ideal by which many more moderate thinkers measure their own views. Berry's highly popular book *The Dream of the Earth,* for example, was published, not by some fringe organization, but by the respected and mainstream Sierra Club. Deep ecology, a far more radical environmental movement that is examined in a later chapter, finds much that is congenial in Berry.[11] Since Berry and his collaborator, Swimme, clearly and competently articulate some deep elements that are common to, but sometimes only implicit in, various branches of religious environmental thought, it is worth taking a closer look at their whole analysis.

A New Beginning

They begin at the beginning of everything. Instead of the old Newtonian universe or even the mere raw data of the new cosmology, they insist that the new universe story is better described by the term *cosmogenesis*.[12] Cosmogenesis points less to a thing, the cosmos, than to the ongoing emergence of the world, from the big bang down to our own time. The universe with which we are familiar must be conceived of as a single, connected process with bonds among all things. At the same time, the process branches off into three main subdivisions: differentiation, autopoiesis, and communion. These terms at first sight may appear forbidding and rather abstract, but, carefully examined, they can be seen to represent some solid and important notions.

Despite the unified unfolding we observe, for example, the universe has clearly *differentiated* itself into an almost infinite number of forms: "At the heart of the universe is an outrageous bias for the novel, for the unfurling of surprise in prodigious dimensions throughout the vast range of existence."[13] Subjectively, this is the source of our delight in the manifold riches of creation that we see, hear, smell, and taste all around us. Differentiation is also a principle of environmental action, suggesting that we should preserve in nature and in human societies all the complexities that the universe has stored within it since its origin. Biodiversity and human diversity may thus come to be seen as part of the one cosmic process.

Autopoiesis in this system refers to the capacity for elements within the larger universe spontaneously to organize themselves into structures of varying levels of complexity. From the first hydrogen atoms that were formed out of free protons and electrons after the cooling of the original fireball down to the development of varied forms of life that seek to preserve their own lives and propagate themselves, we see, say Berry and Swimme, one of the basic patterns of the universe. Looking at the whole universe story, we also need to recognize that there are costs and benefits to be taken into account all along the way. Out of fidelity to scientific truth, Berry and Swimme do not hide the fact that whole stars and worlds were destroyed and tens of millions of species disappeared on earth long before any human being existed here.

Earth, then, was a development with novel features paid for by the destruction of massive astronomical bodies along with, perhaps, planets bearing life. Thus, even the lowliest form of matter we see around us, minerals in rocks, arrived here at great cost and with great potential: "At the very least we can say that the future experience in a latent form is wrapped in the activity of the rocks, for within the turbulence of molten magma, self-organizing powers are evoked that bring forth a new shape — animals capable of being racked with terror or stunned by awe of the very universe out of which they emerged."[14]

Yet, amidst all this cosmic creation, destruction, and reorganization, *communion* remains a basic scientific and human fact. In quantum physics, any particle that has once interacted with another continues to be influenced by it forever. Berry and Swimme assert: "Alienation for a particle is a theoretical impossibility."[15] So, too, do the largescale structures of the universe such as galaxies, nebulae, and stars reflect that same universal interaction. Human beings, by their place in nature, have a similar interconnectedness with all beings, but we also have the capacity to ignore that fact, with unfortunate results: "The loss of relationship, with its consequent alienation, is a kind of supreme evil in the universe. In the religious world this loss was traditionally understood as an ultimate mystery. To be locked up in a private world, to be cut off from intimacy with other beings, to be incapable of entering the joy of mutual presence — such conditions were taken as the essence of damnation."[16] Though the words are notably not used here, the point the authors are clearly trying to make is that human beings may sin, and that finally this turning away from reality toward the isolated self is Hell.

Berry and Swimme situate these three cosmic subprocesses in a well-crafted narration of the universe story that gives them an immediately graspable significance. Swimme provides the scientific depth and Berry the poetry. Sometimes this involves them in jury-rigging early creation myths. For instance, the Middle Eastern serpent-goddess Tiamat, whom we encountered in the first chapter, is the name they give to the supernova that provided our solar system with its heavier elements — a parallel with the Babylonian myth in which the god Marduk slays Tiamat and uses the parts of her body to create heaven and earth. Unlike Joseph Ratzinger, however, Berry and

Swimme are not concerned about the dark dimension in that myth that introduces evil into the very constitution of the universe. The way they use several myths to convey the universe story unnecessarily weighs down good science with other bad associations. But taken on the whole, Berry and Swimme's work retains its value because the story line they create makes clear that the vast spaces and times preceding human existence have an intimate relationship to every nook and cranny of our lives. We are part of that cosmic story even in the most rigorous scientific terms.

A Functional Cosmology

This is particularly important for most people today because, without what the authors call a functional cosmology, all the discoveries of science about the universe's origins and development — and therefore nature itself — seem to alienate us still further from our world. Instead, their vision allows them to join religious and scientific categories in ways that extend their human significance. For example, the cycles of creation–destruction–re-creation in the universe confront all of us with the necessity to sacrifice. In its etymology, *sacrifice* means to make holy something that may be painful or necessary. Sacrifice, they rightly warn, should not be neurotically embraced for its own sake. But seeing effort and suffering as inextricably part of the universe story restores meaning to human experiences that otherwise seem mere pain and emptiness.

They wisely note that failure to acknowledge the need for sacrifice prevents us from achieving human maturity:

> An individual who takes as a central life project the avoidance of pain and suffering of any sort will lead a neurotic and ephemeral life. A society that takes the elimination of all hardship and all suffering as the essential aim of its major institutions will create a flat existence for its humans and a deleterious world for its nonhuman surroundings. And the total cost to be paid in either case is monstrous, enormous, grotesque.[17]

Heroism and self-sacrifice are highly prized in all cultures, and courage in facing pain and difficulty is a central and admirable quality of what it means to be fully — and perhaps even cosmically — human.

But as is also clear in this passage, the authors seem to believe that developed societies have fallen into the trap of wanting to avoid sacrifice at all costs. We seek a technological Wonderworld on the Disneyland model, according to Berry, and are therefore creating a Wasteworld in our flight from limits. Toxic wastes, acid rain, ozone depletion, soil erosion, species loss, and greenhouse threats are the specific forms our Wasteworld has taken. The human race has been locked in a Technozoic era in which we have tried to control everything; the plight of the earth now calls for us to move into an Ecozoic Era.[18] The roots of this development go far back into the human past and call for a profound uprooting of long-standing personal and social attitudes. Like the biblical prophets of old, the cosmogenesist and the geologian tell us we must recognize our ecological sin and repent if we are to escape the destruction to come.

A Different Reading of the Past

Of course there is some truth in each of these charges, but the religious categories in which they come are at least as misleading as they are revealing. Part of the problem is that when Berry and Swimme turn from cosmology to human society, their vision does not serve them particularly well, precisely because they have intentionally abandoned traditional religious categories as allegedly inadequate to their purposes. Berry has become so focused on the religious dimension of ecology that he has gone so far as to recommend that we give up the Bible for twenty years because of the emphasis it places on sin and salvation to the neglect of the goodness of creation.[19] That might spur attention to ecological theory and practice, but there are strong religious reasons to doubt it. We know from accumulated religious wisdom that lives usually change only after a deep confrontation with their own sinfulness and helplessness. To assume that people, untouched by an experience that usually reproduces patterns of salvation in the Bible, will have the insight and the will to be different because they see problems in the human interaction with nature is highly optimistic and impractical. And those of us who must be born, live, suffer, sin, repent, forgive, reconcile, or die over those twenty years while ecologists are busy saving the earth may think we need the wisdom and spiritual power of the salvation story in the meantime.

But this is only the beginning for Berry. He goes on to make sweeping assertions about history, politics, and economics that not only are unfounded but also are often demonstrably wrong and potentially dangerous. In broad outline, he thinks that we were at our best roughly twelve thousand years ago in the Neolithic village and that we should make those villages a model for human communities. A hint of biblical language appears in this scenario. The rise of man coincided with the great proliferation of species to constitute the high point in human experience: "Perhaps the only word to describe the world that gave birth to the human form of life is paradise."[20]

For Berry, however, the transition from hunter-gatherer societies during the Neolithic period to stable agricultural and herding settlements was the first step away from the right human balance with nature. And the rise of urban civilizations around the world beginning around five thousand years ago — in Europe, the Near East, Meso-America, and China — already showed the beginning of an unwieldy divorce from nature. He allows that cities may be necessary:

> A case could be made for keeping the village as the normal context for human development, with the provision that there also be a limited number of larger urban centers where some aspects of the human might find more extensive forms of expression. . . . The difficulty is that the large urban centers have assumed the role of being the ideal context for developing an authentic mode of the human, and that village life is considered essentially as retarded life and an unworthy context for the truly human.[21]

By the time we get to the modern period with its nation-states, capitalism, industry, and science, the decline from the gentle, environmentally benign forms of the Neolithic village to an environmentally rapacious human culture is complete.[22]

On Berry's own principle of cosmogenesis, however, these more complex forms of human organization might alternatively be seen as a continuation of the differentiation, autopoiesis, and communion he sees in the rest of the universe. Where these processes have unintentionally impinged on other important natural processes, perhaps we will want to sit down and think a bit about how to handle the ill effects. But there is a deep impulse running through Berry—and many of his admirers as well—that would like to do something inconceivable in the cosmic process they otherwise celebrate: return to some earlier, highly idealized phase. Not only would that mean death to many of the poorest peoples of the world now precariously holding on to life but also it would create an impoverished human prospect in general. The Neolithic village was an important human setting where it was possible to live a fully human life, and we all owe it a great deal. But, by comparison, a modern village in rural India, even with its high rate of infant mortality, diseases, and general poverty, would seem luxurious. Few of us would choose to return to such an environment.

The Bioregion Project

Along with championing the Neolithic village, Berry and many other environmentalists also advocate what has come to be called bioregionalism. In this view, various natural environmental regions should be identified and organized so as to be self-sustaining. There is an immediate attraction to this vision, as there is to the village. The village restores the face-to-face community that many feel is lacking in large modern societies. And the bioregion project seems to force human beings to exist sustainably within a given natural habitat. All foods and materials would have to be produced locally, and presumably cultures would have to be shielded from outside influences that might give local communities ideas. What is not good for these discrete entities, as some environmental authority would have to determine the good, cannot be good for the larger biosphere.

Critics have pointed out that it is difficult, if not impossible, to identify such natural regions. But the human question is even more complex than the ecological question, and the proposal shows some worrying dimensions once we reflect on them more closely. Should African blacks, for instance, whose hair and skin are clearly adapted to the strong sun of the tropics, live outside of their "natural" regions? Would agricultural specialists, engineers, and investment capitalists in the north have any claim to activity in the south? What begins as a rather nostalgic impulse to create and preserve local communities and practices can degenerate quickly into humanly limiting prospects.

And there are other moral issues that stem from the dogmatically cosmic approach to human communities. For example, Berry deplores the modern nation-state. Instead of seeing such states as at least partly successful attempts to integrate people into viable entities with similar language, customs, and history, or as a means that we have found to manage certain problems and free ourselves from age-old tribal conflicts, privileges, or tyrannies, Berry alleges that states simply pit one group against another. Their former political imperialism has changed into economic and cultural imperialism, he charges. The level of analysis in these passages is closer to a kind of vulgar Marxism than it is to, say, the wisdom and sophistication of the *Federalist Papers*. In Berry, the modern nation-state has led to wars, repression, and jingoism — human evils not exactly unknown prior to the modern age — and modern corporations and electronic media raise serious questions about how to preserve local diversity in the face of the increasing globalization of economics and culture. But the modern state and corporation have also been, by any balanced accounting, great vehicles for basic justice and general prosperity.

At times, Berry casts the whole modern drive to improve the human condition as a kind of demonic force:

> The determination to dominate the universe so that all insecurity, limitation, destruction, and threat of destruction could be eliminated eventuated in racism, militarism, sexism, and anthropocentrism, dysfunctional maneuvers of the human species in its quest to deal with what it regarded as the unacceptable aspects of the universe.[23]

The old modern pursuit of mastery undoubtedly has been the source of many human problems as well as environmental ones. But was it a decline from, or an improvement on, the Neolithic village or other primitive tribal societies that Berry admires? The freedom, prosperity, and security of developed states attract people from all over the world, including many from societies approximating those Berry sees as ideal. Those immigrants might testify that Berry is exaggerating unfortunate features of the developed world at the same time that he minimizes some equally important truths about the alternatives.

Even the evils inherent in the development of the modern nation-state pale in comparison, in Berry's reading of history, with the emergence of the multinational corporation. For him, there are not good

corporations and bad corporations, or even entities that are a mixture
of the two, as we might expect from human experience. The concentration of power in multinational hands, a cross-border unity he desires in other contexts, has created an "industrial myth":

> These new institutions directed vast scientific, technological, fi-
> nancial and bureaucratic powers toward controlling Earth pro-
> cesses for the benefit of human economy. By the end of the twen-
> tieth century, the destruction left by the wars between nations was
> dwarfed in significance by the destruction of the natural systems
> by industrial plunder.[24]

Berry's solution to this "industrial plunder" (human industry and
creativity, however flawed in their myopic execution, and the benefits that have accrued to the human race from them do not appear as
part of the unfolding cosmic process for him) is clearly spelled out in
several places in his work. He wants a "biocentric" rather than democratic control of economic enterprise that will limit us to only those
actions that benefit the entire living community of the earth: "This is
not socialism on the national scale. Nor is it international socialism,
it is planetary socialism."[25] Berry claims to find warrant for such socialism in Saint Thomas Aquinas, though the Angelic Doctor was a
firm believer in the justice and efficiency of private property.[26]

Planetary Socialism?

You do not need to have an inordinate fear of socialism to discern
here a curious and self-contradictory rigidity and no little utopianism. On the one hand, the old scientific era with its techniques and
industry is seen as a human narrowing, a "technological entrance-
ment."[27] Berry even includes the old, existing socialism under the indictment: "Every modern economic system — from the mercantile
and physiocrat theories of the seventeenth and eighteenth centuries,
through the free-enterprise system of Adam Smith, through the socialist economics, to the supply-demand theories of Keynes — is anthropocentric and exploitive in its programs."[28] On the other hand,
he does not seem to perceive that *any* human society, however enlightened by greater awareness of ecological problems, must exploit
and "plunder" the earth merely to exist.

Furthermore, it is hardly news that the record of "real existing socialism" on environmental questions does not give us much ground to suppose that an international or global environmental socialism would be much better for human beings or nature. Democratic societies, partly because of the competitive mix of interests they represent, have demonstrated capacities for self-correction that regimes from more centralized, abstract, perfectionist schemes bred — even Ecozoic ones — have not. Markets allow a degree of flexibility, decentralization, and complexity that Berry praises when he finds it in the cosmos but finds threatening when it is practiced by humans in their democratic politics and economics.

Not every activity in the market is lovely in human or environmental terms, but neither is every natural occurrence. As one market advocate has put it:

> The market is not a place or a person or a conspiracy. It is a process — a continuous adapting system of trial and error, experiment and feedback, freedom and responsibility. It rewards both discipline and risk taking, creativity and deferred gratification, foresight and learning from the past. The market process makes possible not only commercial activity but the countless voluntary associations that arise spontaneously when people are allowed the freedom to form their own bonds. Because it depends not on predetermined status but on contract—on choice and consent—the market is liberating. But it is not, as its critics charge, "atomistic," except in the sense that atoms have a tendency to form molecules, which in turn create larger structures.[29]

None of this praise of markets is inconsistent with concern for the environment, with criticism of socially irresponsible corporations, or, for that matter, with a dynamic and spiritual vision of man and the cosmos. It simply points out the prerequisites — right political and moral order — both to enable and to guide economic activity. By contrast, Berry and Swimme have committed the old intellectual error of thinking that they are able to translate scientific or religious principles directly into human social systems, despite the experience of centuries with the dangers of such totalitarian or theocratic experiments. They have often been the targets of critics of a conservative bent, but even some leftists see Berry and

Swimme's cosmology-driven vision as threatening a new scientific "hegemonic discourse" that may also portend "baleful consequences for the environment."[30]

Remarkably, in the final analysis, Berry and Swimme admit that they do not have much to recommend to us — other than to shun greed, materialism, capitalism, the multinational corporation, and the nation-state. They suggest that we return to greater awareness of our connection to, and reliance on, nature. They ardently call for the inauguration of the Ecozoic Era through the adoption of more earth-friendly technologies. And they forecast a future in which the fate of the human race will hang on the result of the struggle between Ecologists and Entrepreneurs.[31] Yet they cannot help confessing: "We do not know and cannot know the exact shape and form of the Ecozoic era. We cannot know with certainty even what is required of us now."[32] Limits to growth also attract their attention, but here, too, their concern is empty: "Just what these limits are we do not know."[33] (Astonishingly, in the 1990s Berry and Swimme were still praising the Club of Rome's 1972 report *The Limits to Growth,* an almost universally discredited analysis.)[34] These are damaging admissions for thinkers who have elsewhere prophesied imminent environmental disaster without a swift change of heart. Toward what are the pure of heart to turn?

Wishful Thinking

Berry and Swimme unfortunately indulge in a great deal of wishful thinking in the face of some difficult human problems. They counsel listening, like primitive peoples, to the winds and waters, mountains and plains, to get ourselves in touch again with wonder at creation and wisdom in dealing with it. There is certainly nothing wrong with this, and a good deal right, as a means of stimulating our imaginations to take in a wider slice of reality and a way to acquire natural contemplation, which has all but disappeared among mostly urban people. It promises a psychological and spiritual widening of human horizons, but is this a solution to our *environmental* problems? There is almost no evidence of any primitive people living without harm to the land. Their technologies were weak and their effects on nature small, but that was not owing to any lack of will or deep spiritual restraint on the wish to do more. These were, after all, human beings like ourselves.

No one who thinks markets, technology, and corporations have largely been human goods need defend everything these institutions did in the past or may do in the future. But developing nations today, thanks to technologies that have evolved in the advanced countries, will have far less need to engage in the old style of industrial revolution to become more prosperous, healthier — and, to use a loaded word in full consciousness, *better*. Most of these nations seek modern relief from age-old burdens. By contrast, thinkers like Berry and Swimme can only advocate a return to the kinds of conditions that many developing peoples flee at the first opportunity. Developing peoples have seen the Neolithic village. It works tolerably well, and they want to preserve some of its values in their lives. But they also want their future and their children's to be something different.

Perhaps Berry and Swimme are a good illustration of the limits of the cosmological approach itself. Berry wistfully wishes the human race had largely remained in something like the form of the Neolithic village. How we would have had the modern "revelation" of the new cosmology under those circumstances he does not say, nor does he seem very grateful for the advances in agriculture, medicine, and knowledge that have accompanied our departure from the village. But in their own account of cosmogenesis, by what principle can Swimme and Berry say, after fifteen billion years of cosmic development and the rise of humanity, "enough!" All the highest achievements in culture, thought, politics, and religion are ongoing extensions from our humble human beginnings, despite the vacillations and breakdowns in all things human. Human cultures, of course, can take good or bad directions. But the solution for the bad we may find in human cultures is not the essential abandonment of culture per se in the name of some probably illusory prehistoric ideal. Human societies, as they have emerged from the universe story, are complex entities precisely because of all the potential for good and evil that God and the universe have built into us. Respecting and loving the environment on which we rely and of which we are a part are human obligations. But our stumbling growth may be one dimension, and a crucial one at that, of the yearning of the cosmos itself towards greater complexity and a higher life.

Frederick Turner: A Culture of Hope

Far different from Berry and Swimme's pessimistic historiography is the vision of the human future under the new cosmological dispensation that appears in the work of the American poet and essayist Frederick Turner. If the word "visionary" still retains any positive meaning, Frederick Turner must be described as a visionary. Probably not since Goethe has a poet and essayist fruitfully absorbed so much from so many different fields — anthropology, history, physics, mathematics, literature, politics, economics, and theology — while energizing the whole along the graceful arc of his own rich creativity. Turner is a former editor of *Kenyon Review,* a leader in the aesthetic movements known as the New Formalism and the New Narrative, and a cultural figure of astonishing scope. (One of Turner's books, *Genesis,* is an epic poem of several hundred pages about the "terraforming" of Mars, i.e., turning it into a humanly habitable planet like the earth. In the abstract, the subject is not exactly promising, but the poem is vigorous, lyrical, humanly moving, and gripping.) No recent thinker has presented a more energetic and imaginative challenge to so many fashionable intellectual assumptions about nature and human nature.

His *Culture of Hope: A New Birth of the Classical Spirit* (1995), however, is a wide-ranging summation and extension of all his previous work.[35] Turner is quite hopeful about recent human developments, and, while recognizing obvious past and present impacts on the environment in industrial societies, he regards the growth of science and technology as an undeniable expression of essential elements of human nature and the cosmos. He accepts the usual historiography that sees the last three centuries of emphasizing material processes as a detour from a fuller vision, but he gives it an original twist:

> we needed three centuries of self-imposed alienation, of tearing things to pieces to see how they worked, to be able to come back to a coherent world, this time with the powers and knowledge we always felt were our birthright — powers and knowledge we had mimed with magic. But now that we have come back we must cast away the habits of exile — the self-contempt, the illusion of alienation, the hatred of the past, the sterile existentialism, the fear of the future, the willful imposition of meaninglessness on a universe bursting with meaning.[36]

Turner does not, therefore, advocate a retreat to the Neolithic village or to any other idealized point in the past. Instead, he willingly embraces the adventures and perils of the future.

Reversing the human thrust toward increased knowledge and power would be a mistake and a disastrous refusal of our full cosmological destiny. And even trying to slow down change may go against the cosmic grain:

> We are descended in a direct evolutionary line from natural animal species, and are ourselves a natural species. Our nature, certainly optimistic, transformative, activist, and bent on propagating itself, is not unlike that of other species, only more so. We are what nature has always been trying to be, so to speak. Nor can it be objected that it is the *speed* at which we transform ourselves and the world around us that is unnatural. Higher animals evolve faster than more primitive organisms, just as they in turn do so faster than nonliving systems. If we take flexibility, complexity, hierarchical organization, and self-referentiality as the measure, we may define nature as acceleration itself. For us to slow down would, if we take nature *to date* as the model of what is natural, be an unnatural thing to do.[37]

After the widespread caution about our capacities vis-à-vis nature, Turner's general confidence that we not only will manage environmental challenges but also are speeding bearers of cosmic purposes is quite breathtaking.

Though environmental theorists do not fall neatly into familiar ideological groupings, Turner's position is unique even in the current jumble of liberal and conservative thought about ecology. Some of his views have a strong traditional tone, others break entirely new ground. For instance, he argues on the basis of recent anthropology and neuroscience that classical genres in the arts, older and saner social and familial arrangements, and a recognition of goodness, truth, and beauty — the traditional transcendentals — are prescribed by nature as essential to a fully human life. Contrary to those who would deny traditional notions of human nature in the name of freedom, he also contends that hierarchy, norms, standards of excellence, and discipline not only are good in themselves but also are preconditions to true human freedom: "human nature does indeed exist: however, our nature is not a limitation of our freedom but the very source of it."[38]

Yet he is far from being a mere traditionalist. He seeks new forms of the perennial instead of a restoration of the past, and his grounds for doing so will, at times, put off many who might otherwise agree with him. Drawing heavily on the much heralded recent scientific discoveries in chaos and complexity theories, he takes a very different view of our current state and prospects than Thomas Berry. He proposes a "radical center" that would recognize "that value is continually being created by the natural universe, that it is not a shrinking pie, and that human beings can share in and accelerate the growth of value through work which may be delightful, if disciplined. Value of this kind should circulate freely where it is needed."[39] His cosmology has a political and social component as well:

> The present market system of capitalist economies is a linear and clumsy attempt to imitate the more subtle processes of true value-creation, but though it is our closest approximation to date (much closer than any socialist system) it is transforming itself as our technology enables more perfect and multidimensional forms of communication. As machines take over the drudgery, the labor basis of value is being replaced by an information basis of value; and this in turn will be replaced, perhaps, by an emergent kind of value which is hard to define but as a kind of embodied grace.[40]

Far from thinking science and technology are leading us away from nature, then, Turner sees them both as currently guiding us towards a new, and much richer, conception of the great chain of being. Unlike the older notion, the great chain of being underwritten by postmodern science will be dynamic instead of static, will allow for the emergence of new species and forms, will show both bottom-up evolution and top-down guidance at various levels in the natural hierarchy, with God as a creator internal to nature, while nature, properly understood, is self-transcending. Evolution in this light becomes a cosmic process producing greater order, freedom, and meaning— not the mechanical natural selection by random variation as in Darwin. And man stands at the top of this chain as its highest expression and conscious shaper, a position that, while it should induce in us much fear and trembling at our need to use other beings, should also give us hope. We are creatures not merely of desires that drive, but also of hope that uplifts.

Mistaken Environmental Assumptions

Turner believes that our current mistaken ecological framework is the result of several cultural transformations. Once the Newtonian cosmology had become dominant, it appeared as if the only realm of freedom and meaning was human culture. Immanuel Kant was the primary exponent of culture as the alternative to natural necessity. But this compromise, which preserved human things by separating us from the natural world, was inherently unstable. One reaction that set in was the Romantic notion that human culture restricts us, while nature is wild and free. Another developed classical and religious currents to oppose culture, now free but destructive, in the name of nature as order and balance. "Out of these reversals of the traditional dualism came the radical environmental movement."[41]

Technology became the villain of the piece when it was associated with a humanity that was free, but a wrecker of nature. Rejection of technology came from several quarters. Rousseau and his followers posited nature as good and our interventions in nature as evil. Conservatives of various stripes, sometimes anticipating lines of argument that we just encountered in Thomas Berry, rejected technology and industry as supposedly disrupting settled communities, organic complexes, and, for the conservatives, age-old authority. On the political Left, technology was, at first, embraced by Marx as the liberator of the race. But capitalism's superiority to socialism in delivering the goods has turned many on the Left into opponents of technology. In general, environmentalists have embraced a despairing view that technology, the very means by which humans know and shape the world, is unnatural, and environmentalism has thus become both antihuman and pessimistic.

Turner shares some of the concerns of the environmentalists but wants to turn the whole effort to restore and even improve nature in a different direction. The problem for him is not so much environmentalism per se as the wrong assumptions that presently lie behind the typical forms of environmentalism. These, briefly, are that "nature is homeostasis"; that happiness is doing the will of nature; that human happiness is static and change is evil; that we are different from, separate from, and subordinate to nature; that we are no better or more important than other beings; and, finally,

> that an (unelected) community of environmentally conscious, morally refined, sober, devout, humble and self-denying ecological Brahmins should interpret to the masses the will of Nature and direct them accordingly, chastising the merchant/industrial caste, humbling the warrior caste, and disciplining the farmer caste.[42]

As we have already seen, however, many of these assumptions, beginning with the idea that nature is static or self-regulating, are often simply wrong and, even when they are right, need further qualification and refinement. Turner recalls that the geological events that killed off large percentages of species in past eras led to recoveries that were more diverse still in terms of life forms. Only parts of the biosphere can be described as self-regulating and in balance through negative feedback of the kind that enables a thermostat to control the temperature in a house. "Other parts, however, are involved in positive feedback processes that are irreversible, catastrophic to their predecessors, and often wildly original and creative."[43]

Given his belief in ongoing cosmic evolution, Turner simply dismisses the "steady-state" and sustainable-development ecologists as themselves antinature. For him, their whole position relies on thinking that stasis in nature is good and change bad. But the cosmic story does not point to stasis. Man is part of the universe, and to hold back our technological evolution would be to go contrary to everything within and outside us. Early industrial technology, he allows, did create some temporary problems, which are largely on their way to solution. And future technologies will be smaller, more powerful, probably concealed in our environment, something almost like a magical prosthesis that we will be able to summon and dismiss at a thought.

All this may seem to indicate that Turner is not much concerned about specific environmental problems, but he returns to them with a form of the Gaia hypothesis specially crafted to correct mistaken environmental assumptions on the basis of emerging understanding. In a vein similar to the cosmic evolutionist Teilhard de Chardin — and with many of the same virtues and vices as his predecessor — Turner would like us consciously to embrace the thrust of nature toward greater complexity, hierarchy, and sophistication, and to further it in our role as the most self-conscious portion of Gaia. His Gaia includes the divine, which means that science is a kind of theology that gives us knowledge of nature and ourselves. Gaia changes, has a story, is both one and many, and is still only a fetus. We are Gaia's nervous system, and our technology is, and should more deliberately become, a way to serve Gaia; and to serve Gaia is "to increase the scope, power, beauty, and depth of technology."[44] Indeed, he states baldly: Gaia is the process of increasing technology. Ultimately, we are "wiring the brain of God."[45]

A More Progressive Progress

How does this proceed? Turner is unafraid of invoking the old term "progress," which in his view

> is the continuation of the natural evolution of the universe in a new, swifter, and deeper way, through cooperation of human beings with the rest of nature, bringing conscious intention and organized creativity to the aid of natural variation and selection. Evolution in these terms is a spiritual and mysterious process. This evolutionary process is one in which there are honorable roles for technology, for natural conservation, for art, and for economic development.[46]

What Turner restores here is something missing from the timid, half-guilty fears of much environmentalism: a sense that humans are engaged in a remarkable adventure in which they will encounter many challenges and may often fail but may feel an inspiriting hope in the future.

Turner also does not shy away from seeing the human race as the designed culmination, to this point, of the cosmos. For instance, against those who think we need to behave in ways that respect the harmony of the interdependent natural elements in an ecosystem at its "climax," he argues for recognizing the interdependent web as well as the natural hierarchy present: "It is easy to distinguish the low-level, low-energy producers from the more complex, refined, and high-energy herbivores and predators that depend on them, but which also, by their reproductive and consumptive cycles, govern and set the tone for the whole system." And this empirical fact reflects a scientific and metaphysical truth: "In other words, interdependence is not the opposite of hierarchy, but the precondition for the dynamic and nonlinear emergence of natural hierarchy. And natural hierarchy is creative, flexible, free, the very opposite of closed, rigid, repressive."[47]

This seems to embrace every element of hubris with which the typical ecologist charges Western thought, but Turner is not quite so easily pigeonholed. In fact, he goes on the counterattack. He states flatly that while we are in necessary solidarity with nature and may choose some vague harmony in certain contexts, the very idea that we can be in harmony with a dynamic nature is "essentially incoherent."[48] Similarly, the notion that all species are of equal worth and to be respected within some overarching balance of nature may have elements of truth, but it also has large elements of obvious untruth: "It should not really be necessary to argue whether a human being or an AIDS virus is more valuable, but we are forced to such measures by the assertions of some of the more extreme Deep Ecologists."[49]

In Turner the old modern antitheses get overthrown and redefined. Universal structures involve hierarchy, and hierarchy and freedom become for him coordinate terms. It is only through the enormously complex hierarchy of our physical bodies and brains that our mental and spiritual activity becomes possible at all. As every good athlete, scholar, musician, or dancer knows, greater freedom and unprecedented breakthroughs occur after long discipline, not by following immediate impulse. Nor does fruitful hierarchy stop with the individual: "The more complex and multi-leveled the hierarchy, the greater the opportunity for individuated behavior, free decision, and creative innovation. Similarly, close study of democratic and economic organizations reveals that the more freedom and opportunity and self-expression they permit, the more complex and multi-leveled the hierarchy of function."[50]

In this light, traditional family forms, too, lead more often to liberation than to what has been construed as oppression. Intact families the world over tend to produce freer, healthier, more creative children, while the absence of those familial hierarchies leads with great predictability to passivity, dependency, crime, and drug use. This, of course, contradicts just about everything that has been stylish in intellectual and feminist circles for decades. Freud taught us how our families may warp us, but Freud, who had not seen the modern urban underclass, had no idea how warped people could become without families. Sex itself is transformed in Turner's new chain of being. If the universe bears within itself a creative, free, expansive energy, sex becomes a subdivision or, Turner suggests, perhaps even a sublimation of the spiritual.

For Turner, eco-feminism, which is treated more fully in chapter 6, looks like a natural alliance between ecologists and the women's movement. But it reveals the incoherence at the heart of much current environmental thought: "Both have a satisfyingly contrarian flavor, both seem designed to annoy the imagined world of cold (male, capitalist, scientific) efficiency, both have a warm, pacifist, and emotional tone, both have an egalitarian basis (equality between sexes, equality between species)."[51] But there are several intellectually fatal inconsistencies in this view.

The drive for greater political and social equality, the earlier part of feminism, was only a quest for justice — once modern technology made childbearing and especially breast-feeding less of a burden on women. Earlier, more gender-equal organizations of society would have been impossible. (Turner's father was the great American anthropologist Victor Turner, and the son draws on the father's legacy in these arguments.) The notion that we moved from a period in the past when the Great Mother presided over peaceful and egalitarian human communities to societies ruled by the rapacious male deities of historical record is a paranoid fantasy. Primate studies have shown how possessive, hierarchical, war prone, and sexually brutal are our nearest ancestors, and early humans were probably similar. It is ironic that the radical feminists, who owe their liberation to technology, have made common cause with the radical ecologists in the belief that modern societies are a decline from a feminist ideal. Both have created the myth of the white male who is, according to the myth, angry and aggressive while at the same time he is emotionless and coldly rational.

Moving deftly amidst massive modern problems, Turner traces this and similar confusions to the postmodern bugbear of social constructionism. In the radical constructionist view, we construct the world socially through language and therefore may make it as we wish. In the new great chain of being, Turner says, such assertions will be seen to be nonsense. He concedes that constructionism of a sort is always partly true and, in limited circumstances, may even be absolutely true. But to make our construction of the world always dominant is to neglect how the rest of the world constructs us in ways that both condition and, when we respect them, enable our freedom. He describes the following thought experiment: "When a social constructionist jumps off the top of a tall building, his human construction is entirely outvoted by the mass of the planet and of his own body, and he must submit to rules not made by human beings and fall down, not up."[52]

How can any of us, he asks, accept the feminist and constructionist assumption that in reality there is radical equality in nature that has been disturbed only by our evil search for power? How can any of us continue to coexist without militaristic immune systems and central-nervous-system controls? The vision of a nonviolent relationship to nature is simply untenable: "We already kill every day in a thousand ways in order to live at all."[53] We cannot even look to a so-called sustainable relationship to nature, because nature itself changes cumulatively. Turner takes note of worry over the possibility of global warming only to dismiss it as biased: "It is only the faulty assumption that any change is unnatural that makes us assume that the greenhouse effect will be bad for the planet."[54]

Some Demurrals

This view, and the optimistic orientation toward technology that supports it, would be easy to satirize. And Turner's linguistic gifts and fertile imagination do not always serve him well when he is contemplating the future. He blithely accepts the notion that we may someday engineer ourselves to have gills or fins, as if such disruptions of "natural" evolution would not deny some of the deep structures and hierarchies he has elsewhere championed. He himself makes some questionable assumptions — for instance, that the warmer world produced by greenhouse gases could be like several good winters, "greener, wetter, more fertile."[55] Perhaps so, but we have no greater guarantee of that than we do that global warming would be catastrophic. And should we really be comforted about, say, the ozone problem because "far more catastrophic *natural* changes" have occurred in the atmosphere? This is factually true, but it hardly settles the issue.

At the furthest limits of his vision, Turner seems too optimistic and in too much of a rush to affirm whatever seems to be coming into view. The imbalance here has something to do with his jettisoning of more orthodox theological views. For instance, his whole notion of a culture of hope is based, not on a biblical affirmation of a world to come and a confidence that we will receive enough assistance to get through our daily difficulties, but in a view that makes the cosmic process itself into hope. Turner redefines the old theological virtues: "if faith is the affirmation of what was, and love the affirmation of what is, then hope is the affirmation of what is to come."[56] But faith, hope, and love each point in many directions. Faith in what we have already been given in the universe, for example, is by Turner's own showing the basis for understanding the form that our hopeful steps into the future require. And love is not only of the present but also of what we have from the past and wish to bring into being in the future.

Turner has consciously opted for a God who is not both transcendent and immanent, as in traditional theology, but who is part of nature and evolving with it. This involves Turner, as it does the process philosophers to whom he is akin, in envisioning a God who needs to change and who regularly grows new parts and kills off old ones. The biblical God is a mystery, but he is plain as day compared with the God who is part of the universe. Much of what Turner has to say is compatible with more traditional views of a God who enters human and cosmic history in ways also compatible with the new cosmology. But his readiness to imagine radically altered human beings and whole planets hints that in the end there may be too much change and openness and not enough stability and solidity in Turner's cosmos.

His remarks on religion can sound overly sweeping and simplistic: "My claim is that nature itself, like ourselves, is fallen, is falling, and has always been falling, outward into the future from the initial explosion of the Big Bang; onward into more and more conscious, beautiful, tragic, complex, and conflicted forms of existence, away from the divine simplicities and stupor of the primal energy field."[57] This is clear enough, but it is a strangely breezy presentation of a

progress into the tragic, complex, and conflicted. There appears here
something usually absent from Turner's otherwise vibrant vision: a
poetic tendency to celebrate things as they are for their grandeur and
impressiveness instead of a grappling directly with concrete issues of
better and worse, good and evil, fallen and redeemed.

Instead of suffering and sin, Turner strongly emphasizes the need to
recover the notion of shame. Shame for Turner means the feeling we
have at our weakness and dependency on sex and on other living things
for reproduction and nourishment, on other people and social structures
for psychological and spiritual development. The wish to live a purely
individualistic life without such "dependencies" has led us to seek impossible purity either as the radically autonomous modern individual or
as the vegetarian/ecologist who does not wish to disturb the order of the
universe or to live at the expense of other beings. Turner rightly pro-
nounces these wishes impossible and recommends instead that we recognize at what great cost we live and seek to live worthily. For all his
technological optimism, Turner acknowledges that the human choices
between good and evil, the possibility of tragedy, and the paradoxes of
life will never go away entirely.

Yet his strong case for ever-acclerating development bears other
dangers. Both our biblical and civic humanist heritages tell us that in
affluence men tend to grow less virtuous. There is, of course, no law
of human history that dictates that this must always happen. But sta-
tistically, when the necessities of life are easily available, not only do
people take for granted that material comforts can be obtained without effort, but they begin to view them as a right. Shame has evaporated in most developed societies; in the technological world Turner
envisions, shame may become an utterly unintelligible idea, and its
absence may lead to the even further flattening of human existence
that we have experienced since the Enlightenment.

Nonetheless, the mere fact that Frederick Turner can start us
thinking, deeply and systematically, about so many disparate human
issues is a sign of a great synthesizing and visionary mind at work. He
speaks of the chivalric morality we might learn from our domestic
animals, if we had eyes to see; of transcultural practices of asceticism
and self-discipline that need to be adapted to our situation; of the
specialness of the student-teacher relationship and how it should in-
struct us about the fear of being told what to think; of the imperial

incursions of literary theorists; of the need for meaningful rituals to reflect our human creatureliness; of the ontological status of life as something infinitely more complex than mere mechanism.

As a poet, finally, he makes a plea for beauty as one of the three transcendentals. We have tried too long to live, he says, on truth and morals. I would argue that scientific truth has been much with us, but, *contra* Turner, moral truth has all but evaporated in the modern world. Yet he is correct that we need the expanded vision of beauty, too, to make us whole. And what is beauty? It is "a recognition of the deepest tendency or theme of the universe as a whole."[58] And what is that theme? Unity in multiplicity, complexity within simplicity, creation, rhythm, hierarchy, and self-similarity. The scientific notion of the iterative feedback principle, which describes the infinitely complex operations of our brains and of other complex and chaotic systems in nature, is the "eros and logos of Nature." Humans and human arts fit snugly into the cosmos: "We have a nature; that nature is cultural; that culture is classical."[59]

Not only our arts and humanities but also our politics and economics need to recognize that "the highest forms of profit are designated by the terms truth, beauty, and goodness."[60] It is characteristic of Frederick Turner that in that ultimate plea for the transcendentals he is not reluctant to invoke the much despised market and the dreaded word "profit." Contrary to the avant-garde artist who seeks a purity uncontaminated by the sordid commerce on which his grants and subsidies depend, Frederick Turner wants to transform the market by reminding its members of what they already know: that truth, beauty, and goodness have cash value, as William James put it, and value beyond anything else in the market. Not every one of Turner's analyses will survive meticulous scrutiny, and some of his recommendations will fall in the actual wear and tear of life. But it is his great achievement that, after reading him, you come away feeling that you live in a different, better, more wonderful world than you thought, and that despite our many bewilderments, we are a species with many good reasons for hope.

Making a Choice

Are we forced, then, to choose between the views of Thomas Berry and those of Frederick Turner? Do we live in a cosmos whose fifteen-billion-year history should make us fearful of destroying it by any further human advance, or are our own social processes so at one with the rest of the cosmos that they are a brilliant expression and further development of it? For guidance on this issue, we might turn to the great twentieth-century theologian Romano Guardini.

As his name suggests, Guardini was Italian by birth, but he grew up in Germany in the early part of this century. He was also educated there and grew accustomed to the rapid and vast development of German industry. Yet when he periodically returned to Italy to visit relatives, he was distressed at seeing industrial development disrupting an Italian landscape that had been softened and humanized by thousands of years of human care. At first, he thought technology and industry had to be stopped before they destroyed something valuable and irreplaceable. But then he saw further into the question.

Given that so many human goods come out of our modern developments, Guardini came to think that we needed to shape our modern technologies to fit into the landscape as beautifully as the old vine dressers, stonecutters, house builders, and others through the centuries had inserted their creations into the countryside. This approach has the advantage of respecting both the natural inheritance and the human creative spirit. Obviously, not every landscape can be so preserved, given the nature of certain human activities. But neither were they all so preserved over the centuries. Certain trade-offs were made that still allowed basically healthy and pleasing habitats for human and nonhuman nature.

In fact, Europe, the home of the industrial revolution, science, technology, and all the forces most ecologists decry, is in many places a balanced, green, productive continent today. Even Father Berry has recognized that the soils of Europe in this century are, by common agreement, richer than they were originally. More remains to de done in cleaning up the older effects of dirty industries and in preventing new damage. But there is no question that we can envision a day in the not too distant future when we may have both the humanly pleasing habitats that many environmentalists nostalgically associate with one or another point in the past and the exciting and ever changing adventure of human creation and discovery. All it takes is the will and the skill to recognize that even if we make some errors along this path, the powers the cosmos has put into our hands can be

4

Deep, Deeper, Deepest

As the views of Thomas Berry and Frederick Turner suggest, there is no direct connection between the cosmological views they share and a program for the environment. In Berry, the creative thrust of the cosmos dictates human retreat, lest inestimable riches be lost. In Turner, that cosmic creativity should be embraced and furthered by human creativity. Berry's cautiousness derives only partly from his Catholic beliefs; Turner's exuberance is only partly the result of his vocation as a poet without any easily identifiable religious dimension. Their differences suggest that, even when there is agreement about the deep structures of the natural world, other assumptions about human life usually determine our orientation towards the environment. Some may stem more or less directly from deep principles, others may be legitimately derived (though not strictly deduced) from different theologies, still others depend upon a kind of parallel vision of nature drawn from scientific or ecological sources that may coexist in greater or lesser tension with theological traditions. Disputes about which elements should be paramount, and on which issues, abound. On the one hand, do views of nature that proceed *from* religion, whether to criticize or bolster environmental stances, deserve precedence? Or, on the other hand, should claims about nature that some people believe should be applied *to* religion for purposes of theological reform be our guide? This set of questions may be fairly said to lie at the heart of many current religious debates over environmentalism.

Though many approaches to these questions involve complicated theologies and philosophies of nature that are too abstruse for our

145

purposes, it is essential to look into how religious belief and environmental concern interact. Doing so will enable us to see that there is no simple translation of concerns back and forth between environmentalism and religion. Although it often appears at a policy level that environmentalism has become a new war of religion in which dogmatism drives the battle, with little hope of reconciliation between the parties, at a slightly deeper level there is much more nuance and complexity. In some instances, current environmental concerns may put some tough questions to religious traditions that have never squarely faced such issues. But in others, religious traditions may uncover some blind spots in environmentalism and prevent environmentalists from adopting positions that rely too much on personal impulses rooted in contemporary worldviews and too little on the steadying guidance of some of the accumulated wisdom of the race.

Probably the movement most known for its claims to encourage exchange between religion and ecological practice is Deep Ecology. Founded by the Norwegian philosopher Arne Naess, Deep Ecology claims many ancestors; one list includes Thoreau, John Muir, D. H. Lawrence, Robinson Jeffers, Aldous Huxley, George Orwell, Theodore Roszak, Lewis Mumford, Taoism, Saint Francis of Assisi, the eighteenth-century Romantics, Spinoza, Zen Buddhism, Alan Watts, Gary Snyder, Aldo Leopold, Rachel Carson, and more recently David Brower, Paul Ehrlich, and Lynn White, Jr.[1] It is unlikely that every person in this list would be flattered by the inclusion. And whether Deep Ecology delivers on the promises it makes in the name of such illustrious predecessors remains to be seen. But the invocation of these figures at least indicates the wide historical and intellectual reach claimed by Deep Ecology.

In Search of Common Ground

One of Naess's primary aims was to create a common center for a worldwide ecology network that might draw on otherwise different groups such as Buddhists, Christians, and philosophers of one type or another sympathetic to environmentalism. Naess is the ideal figure for this role because he combines some of the charismatic features of a religious founder with the flexibility of a coalition leader.

Deep Ecology as a movement has both won many adherents and drawn sharp criticism, but Deep Ecology as an idea has come to dominate much religious thought on the environment, whether the thinkers are aware of the influence and whether they describe themselves as Deep Ecologists or not.

A brief sketch of Naess's life will help us to understand the movement he founded. Naess grew up in a fatherless home in Norway and early came to feel nature as a kind of surrogate father. As a child, he had several experiences of the intrinsic worth of living and nonliving things and the sheer wonder of tiny living organisms. When he speaks of the "wide identification" with all of nature that must inform a new ecological sensibility, he uses religious or philosophical traditions as seems appropriate, but he has denied that such identification ultimately depends on sophisticated, mystical, or esoteric phenomena. Instead, all wide identification requires is "the ordinary sensitivity of a loving child."[2]

As a university student Naess turned to the study of philosophy and immediately displayed great gifts in a demanding discipline. In 1939, at the age of twenty-seven, he was appointed to the chair in philosophy at the University of Oslo — the youngest person ever to occupy that post. Specializing in the philosophy of science, modern thought, and — a special love — the work of Spinoza, Naess headed something like a philosophical revival in Norway. He acquired an international reputation as a philosopher even prior to his work in ecology. At the same time, he maintained a love of the outdoors, hiking and mountaineering in Norway, and was the first person to climb several peaks in India. But it was in 1972, after a period of studying Gandhi and contemplating environmental problems, that Naess began to put forward his view of the deep shift in sensibility needed to remedy what he thought of as the crisis in the relationship between human beings and nature.[3]

Deep Ecology, Naess has argued, is not completely opposed to what he calls "shallow" ecology. The latter looks at environmental problems piecemeal or at an immediately practical level and seeks solutions, either by changing technologies to be less harmful or by reforming human behavior with measures like recycling. Naess favors every discrete attempt at this level to reduce the human impact on nature. But for him these instrumental modes are insufficient to address the real environmental problem. Furthermore, they conceal the deeper connection between human beings and all other beings that we must recover if we wish to arrive at a true notion of our place in nature and how we should act. This will entail a shift away from the "anthropocentric" view of nature as merely a storehouse of resources for human use to an "ecocentric" view of nature as an intrinsically valuable system of which humans are both a part and—in some ways —stewards. The depth in Deep Ecology consists in making people aware of a profound relationship that already exists.

The Deep Ecology Platform

Naess has said several times that when once you ask people practicing shallow ecology about their attitudes towards larger questions, you almost always find that they share the Deep Ecology principles without knowing it.[4] Drawing on Spinoza, Whitehead, Gandhi, and some currents in Christianity and Buddhism, Naess tries to show how ultimate religious and philosophical premises and "ecosophies" (ecological wisdoms), despite their differences at the very deepest levels, come together at a slightly less theoretical, slightly more practical level. While the two were camping in Death Valley in 1984, Naess and his follower George Sessions sketched out an eight-point Deep Ecology Platform (DEP) as a point of departure for discussions of facts and norms that should lead to concrete action. He also elaborated the "Apron Diagram" (see figure 1) to show how the DEP is situated vis-à-vis ultimate commitments that converge on the platform as well as the diverse positions and actions that may flow from the platform. All of these elements have become so influential in current ecological thought that it is worth reproducing the DEP in full here:

   1. The well-being and flourishing of human and nonhuman life on Earth have value in themselves (synonyms: intrinsic value, inherent worth). These values are independent of the usefulness of the nonhuman world for human purposes.
   2. Richness and diversity of life forms contribute to the realization of these values and are also values in themselves.
   3. Humans have no right to reduce this richness and diversity except to satisfy vital needs.

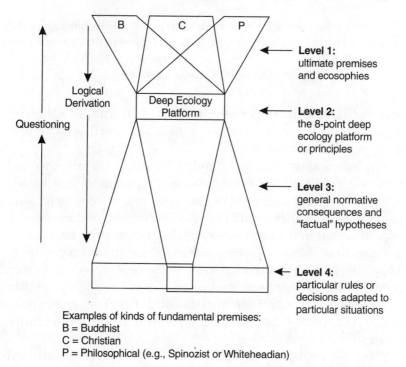

Logical
Derivation

Questioning

B   C   P

Deep Ecology
Platform

Level 1:
ultimate premises
and ecosophies

Level 2:
the 8-point deep
ecology platform
or principles

Level 3:
general normative
consequences and
"factual" hypotheses

Level 4:
particular rules or
decisions adapted to
particular situations

Examples of kinds of fundamental premises:
B = Buddhist
C = Christian
P = Philosophical (e.g., Spinozist or Whiteheadian)

FIGURE 1. The "Apron Diagram"

Reprinted by permission from "The Deep Ecological Movement: Some Philosophical Aspects," by Arne Naess, *Philosophical Inquiry* 8, nos. 1-2 (1986).

4. The flourishing of human life and cultures is compatible with a substantially smaller human population. The flourishing of nonhuman life *requires* a smaller human population.

5. Present human interference with the nonhuman world is excessive, and the situation is rapidly worsening.

6. Policies must therefore be changed. These policies affect basic economic, technological, and ideological structures. The resulting state of affairs will be deeply different from the present.

7. The ideological change will be mainly that of appreciating life quality (dwelling in situations of inherent value) rather than adhering to an increasingly higher standard of living. There will be a profound awareness of the difference between bigness and greatness.

8. Those who subscribe to the foregoing points have an obligation directly or indirectly to try to implement the necessary
changes.

Think like a Mountain![5]

Many of these points may seem to border on truisms. Of course,
there are other judgments in this common platform that may repel as
many people as they attract.

The concluding phrase "Think like a Mountain!" was Aldo
Leopold's cry in *A Sand County Almanac*. As beautiful and loving a
book as that work is, the phrase is meaningless. Mountains don't
think. If they did, they would probably speak with indifference of the
wide variations in flora and fauna that they witness over the course of
geologic time. Mountains exist naked in a vacuum on the moon.
They exist at the bottom of the ocean. The only thoughts any mind
anywhere thinks about maintaining the natural richness and beauty
of a wooded mountain are human thoughts. For all its appearance of
being ecocentric, "Think like a Mountain!" is inescapably anthropocentric.

The same might be said of the first three points in the platform.
We shall examine below the problems in Naess's Buddhism and
Spinozism that make the idea of the equal value of all things incoherent. Here it suffices to say that despite many attempts, no philosopher or theologian has given a convincing argument for the inherent
value of nonhuman nature. The Bible often speaks of creatures as
glorifying God, but that they have value absent their divine connection — an article of faith for many environmentalists — is difficult to
prove. Also, since competing claims must inevitably arise between
various creatures, even if they have value in themselves, we want to
know, other than their utility to us, what their relative value to other
creatures is. If no creature is higher or lower, the Deep Ecologist is in
a bind. For I may choose to value biodiversity for various reasons;
most of the human race does. But if all beings are really on an equal
footing, then if it's a choice between my children or someone else's
father or perhaps even a family dog, on the one hand, and many other
lovely but distant creatures, on the other, it does not matter *at the very
deepest level* which I choose. Point 3 about our having no right to reduce biodiversity, under the equality-of-creatures scheme, appears to

be nothing more than one preferred outcome among many, all equally valid, and a somewhat self-indulgent option for those who can afford the luxury of climbing mountains in Nepal and camping in Death Valley without needing to worry about their next meal.

Point 4 treads on the delicate issue of population and reproduction. As we have already seen, it is far from self-evident that the earth is overpopulated, let alone that "The flourishing of nonhuman life *requires* a smaller human population." Point 5's contention that our situation is rapidly worsening ignores the vast strides already made and being made every day to remedy virtually all environmental problems. Point 6's contention that our basic social systems will need to undergo profound change suggests a drive towards social radicalism, based on alleged environmental trends. And the kinds of radicals Deep Ecology has attracted, despite the often anodyne philosophical formulations by Naess, give no little cause for concern.

For example, George Sessions, the coauthor of the DEP, has argued that "an ecologically harmonious and social paradigm shift is going to require a *total* reorientation of the thrust of Western culture."[6] Other Deep Ecologists, manifesting a recklessness about human consequences all too evident in the movement, have gone so far as to wish for a global economic collapse that would force the West to rethink the industrial system.[7] Ironically, this desire appears in a book entitled *The Arrogance of Humanism.* Such blithe invitations to catastrophe warn us that the shift from the anthropocentric to the ecocentric may be accompanied by a good bit of ruthlessness and arrogance of its own. These confident desires for anti-Western and antimodern revolutions contrast sharply with Naess's usually more careful and pluralistic approach. But Naess himself and the whole movement have been tarred with the brush of fanaticism by those who rightly see many zealous Deep Ecologists flirting with what critics have called ecofascism.

Ecofascism and Ecotage

Some of the outrageous positions in this quarter should serve as a warning. Deep Ecology has been embraced, for example, by the radical group Earth First!, which has not balked at committing "ecotage"—direct sabotage of equipment and sometimes dangerous practices like spiking trees, making them dangerous to cut down or pass through mills. Such practices clearly suggest that some Deep Ecologists believe that they can ride roughshod over human concerns because of their sense of self-righteousness about human incursions into nature. Al Gore, though often given to sweeping claims himself, sensing the brutally radical impulses within Deep Ecologists, felt it necessary to distance himself from the antihuman dimensions of statements like the following by Earth First! cofounder Dave Foreman: "It's time for a warrior society to rise up out of the earth and throw itself against the human pox that's ravaging this precious, beautiful planet." Mike Roselle, another Earth First! leader went further: "You hear a lot about the death of nature and it's true, but nature will be able to reconstitute itself once the top of the food chain is lopped off — meaning us."[8] Some Deep Ecologists have openly hoped for AIDS to reduce the human populations. And even the often sainted Lynn White Jr. once remarked in an article ironically entitled "The Future of Compassion" that we should "light candles before the saints requesting a new Black Death" to ease our ecological peril.[9]

Because of criticism from Gore and others, some of the central figures in Deep Ecology have claimed that casual violence on behalf of the earth has never been part of their program. Sessions criticizes what he calls Gore's "species chauvinism" even as he argues that misanthropy should not be attributed to the movement.[10] In a series of replies to points raised by critics, Naess remarked that perhaps nonviolence should be added to the eight points of the Apron Diagram. To charges that Deep Ecology worries only about species, not individuals, he suggests that perhaps protection of individuals is a point that might be added, too.[11] Yet for all the protests, Deep Ecology sets loose some impure spirits that mere verbal tinkering with the platform may be powerless to restrain.

Though Naess is normally careful to avoid alienating potential allies, even he occasionally says things that would give a thoughtful person pause. For example, in a 1982 interview he presented what he thinks of as the alternatives in a way that could be taken as a kind of blackmail:

> Within fifty years, either we will need a dictatorship to save what is left of the diversity of life forms, or we will have a shift of values. A

shift of our total view such that no dictatorship will be needed. It is thoroughly natural to stop dominating, exploiting, and destroying the planet. A "smooth" way, involving harmonious living with nature, or a "rough" way, involving dictatorship and coercion — those are the options.[12]

For a man who allegedly harbors no sinister political designs, this sounds curiously close to the submit-or-else school of political reeducation.

For another central figure in Deep Ecology, even more inflammatory language is not beyond the pale. Bill Devall, a longtime collaborator with George Sessions, has suggested that there is a moral equivalence between the Holocaust and damage to nature. Devall has described students of natural resource management as akin to "guards in Nazi death camps. . . . their neutrality towards forests or wildlife or fish kills any natural feelings of empathy or sympathy."[13] Critics may have reacted too strongly to this passage in isolation from other elements in Devall's work. But it is not unreasonable to worry that someone who believes that even those who care enough about nature to study how to preserve it are unconscious monsters may be harboring a kind of rage belied by his other words. Such suggestions are not infrequent among certain Deep Ecologists.

Gore was right, then, to pounce on Earth First! and, by implication, a whole swath of card-carrying or fellow-traveling Deep Ecologists. But he is wrong in suggesting that Arne Naess and Deep Ecologists more generally "seem to define human beings as an alien presence on the earth." "Define" is the operative word here. Insofar as there is a clear formulation, Naess and Deep Ecology cannot, on their own principles, define any living being as an alien presence. Gore's further charge that Deep Ecologists do not believe in free will is further contradicted by many of Naess's statements about the special relationship that humans, given their unique powers, have to the rest of nature: "The richness of reality is becoming even richer through our specific human endowments; we are the first kind of living beings we know of which have the potentialities of living in community with other living beings."[14] Unfortunately, that kind of sanity seems insufficient to curb the "ecowarrior" mentality that drives many Deep Ecologists. It is no accident, for instance, that Greenpeace named one of its eco-assault ships *The Rainbow Warrior*.

The best theorists of radical ecology are quite aware of the potential for fascism lurking within overly romantic notions about human beings in nature. As ecophilosopher Michael E. Zimmerman has warned, some attempts at identification with the immortal forces of nature, like the Nazi movement, have led to disaster: "This movement, with its goal of achieving an immanent harmony of *Volk* and nature, proved particularly attractive to those who shunned the heavy moral responsibilities conferred by the Jewish and Christian traditions, according to which humans are not merely 'natural' creatures, but instead exist within a history that is tied to the transcendent."[15] Particularly in the 1960s, diverse groups of people cast off moral and religious restraints in the name of a freer and more "natural" lifestyle that generally led to less freedom and even more unnatural societies. Many radical critics have indicted traditional religion for its contribution to human hubris toward nature and its oppressiveness toward human self-expression. But some of the means by which people have sought to throw off the constraints of tradition have also thrown out qualifications and moral complications that would be necessary to any critical reading of the Deep Ecology Platform.

Ecosophy T

As we have seen, Naess distinguishes from the platform itself the various ultimate commitments (Christian, Buddhist, philosophical) that inform the DEP. His views, therefore, are brought into better focus by an examinination of his own personal "ecosophy," a position that he does not expect to be common to all Deep Ecologists and to which he has given the name Ecosophy T. The "T" refers to Tvergastein, a type of quartz found in Norway near the mountain plateau of Hardangervidda, where Naess has built a hut that has been described as "Norway's highest privately owned dwelling."[16] Naess spends a good part of every year in this retreat, living on as little as possible and trying to rest in communion with the mountains.

Asceticism constitutes a large part of his life there. Naess has warned ecologists who hypocritically live quite affluent lifestyles that "ordinary people show a good deal of skepticism toward verbally expressed values which are not expressed in the life style of the propagandist."[17] Around his mountain hut, he asks visitors to step as much as possible on rocks rather than plants; he even fertilizes the thin mountain vegetation with the contents of the toilets. Human beings self-consciously tip-toeing from rock to rock do not present a very exuberant vision of man on earth. But to read his disciples' adoring descriptions of his simple life in the mountains is to encounter the same atmosphere as may be found in some accounts of the life of Gandhi. As with Gandhi, a host of questions arise. One of Gandhi's followers once observed that "it is costing us a fortune for him to live so simply." Behind the simple life on a Norwegian mountain lie many cultural achievements that make the experience something richer than mere asceticism.

Naess reads, cooks and heats with portable gas devices, and even listens to old vinyl recordings of classical music on a simple record player. In sum, he creates a kind of secular retreat in which the typical modern intellectual can enjoy modern conveniences still more by rediscovering their value in simplified circumstances There is nothing wrong with rediscovering such appreciation outside the glut of consumer products that modern life makes abundantly available. Naess, unlike some of his disciples, does not claim that Ecosophy T will mean the abandonment of human thought and achievements to go back to the level of hunter-gatherers or Neolithic agricultural villages. He is at least open to the notion that human decision-making should take place within the context of ecological sensibility, which should leave room not only for different methods of conceptualizing and approaching nature but also for a diversity of solutions to problems.

Naess has stated that "Ecosophy T has only one ultimate norm: 'Self-realization!'" But he immediately adds, recognizing the danger of misunderstanding:

> I do not use this expression in any narrow, individualistic sense. I want to give it an expanded meaning based on the distinction between a large comprehensive Self and narrow egotistic self as conceived of in certain Eastern traditions of [the cosmic self] *atman*. This large comprehensive Self (with a capital "S") embraces all the life forms on the planet (and elsewhere?) together with their individual selves (*jivas*). If I were to express this norm in a few words, I would say: "Maximize (long-range, universal) Self-realization!"[18]

Self-realization in Ecosophy T represents full human maturation and, therefore, identification of self and cosmos.

Buddhism and Spinoza

Part of the inspiration for Naess's Ecosophy T is in Mahayana Buddhism, one of the three explicit ultimate systems in the Apron Diagram. The remark about *atman* in the passage above refers to the Buddhist belief in nonduality, that is, an ultimate oneness of the self with the entire universe. Though the universe we see is passing away and contains much suffering, that cosmic self, variously conceived in different schools of Buddhism, remains. Our everyday consciousness of separateness from nature and others is, therefore, an illusion to be overcome. The result will be a joyous liberation and compassionate identification with all beings.

Another part of Naess's personal ecosophy stems from Spinoza, whom, it will be recalled, he began reading early in his life as a student of philosophy. Writing around the time that Descartes and other central modern philosophers were separating subject and object, fact and value, Spinoza took a somewhat different approach. In his philosophy, he developed some Cartesian intuitions in a direction that closes the gap between human consciousness and the universe by positing the unity of all things within a network of God/Nature *(Deus sive Natura)*. Spinoza was of Jewish background and had studied certain Jewish mystics before encountering Descartes. And his speculations, although so unorthodox that they eventually lead to his excommunication from the Jewish community in Amsterdam, have so powerfully attracted later thinkers that the French philosopher Henri Bergson once remarked: "Every philosopher has two philosophies: his own and Spinoza's."[19]

Though Spinoza retains from his Jewish background what appears to be a religious vocabulary, as in the identification of God and nature, that tradition used those terms in very different ways. Spinoza does not believe in a personal transcendent God; in his view, the Bible uses pictorial language to convey morals. True philosophy is hardheaded and rational, starting with the divine substance. Everything is necessarily a mode of God, even if finite minds are incapable of grasping that process. But the chain of real beings is determined; unlike the biblical view, creation is not free, events are never contingent, and human acts seem free only because we are ignorant of their causes. Theism and materialism are brought closer in Spinoza than ever before.

Spinoza is a complex and great thinker whose ideas Naess adapts to elaborate the wide identification of ourselves with all of nature that undergirds Ecosophy T. But there are many problems with Spinozism, beginning with the way it thinks about human beings. Naess has characterized Spinoza's thought as maintaining that

> all particular things are expressions of God, through all of them God acts. There is no hierarchy. There is no purpose, no final causes such that one can say that the "lower" exist for the sake of the "higher." There is an ontological democracy or equalitarianism which, incidentally, greatly offended his contemporaries, but of which ecology makes us more tolerant today.[20]

Taken literally, this notion might lead to simple human paralysis in the face of God's acting. As with all forms of pantheism, Naess needs to free himself from some of the strict consequences of the doctrine. So he allows that we may eat other beings from need, even though we are not "higher" than they are.

But what else may we do? Here all lines become fuzzy. Can we cut trees to build houses, pave over pathways for travel, drain wetlands that are noxious to human existence? Depending on the given individual, who is after all presumably part of the same process through which God is working everywhere, some of us may choose to jump between the rocks with Naess, others may choose to shuttle between Manhattan skyscrapers in taxis. The identification of God with nature here reaches its logical conclusion. If there is truly no purpose other than what happens, human purposes may — paradoxically — become all that remains, for good and ill.

As several critics have pointed out, for environmental issues, even the kind of Spinozist philosophy that sees everything as interrelated may be wrong in light of new scientific evidence, particularly in population biology, that ecosystems result from the sum of individual beings and their acts rather than organic interconnections. From another quarter, Spinozism has been attacked for submerging individuals in monolithic networks.[21] Perhaps even more important, however, Spinoza, like many of the past Western figures criticized by Deep Ecology, was himself no Deep Ecologist. He shares the linear rationality that is usually criticized when found in his contemporary Descartes. Perhaps for that reason, he took the typical view of animals in the seventeenth century: they could be used for human needs, whatever place they occupied in his philosophical system. Limitations like these, along with similar problems in historical forms of Buddhism, make it clear that whatever ultimate sources ecosophy theory and practice may draw on, they are greatly adapted for the sake of some predetermined ecological goals.

Radical Egalitarianism

Lying behind Deep Ecology's seeming openness and pluralism of views lies a monolithic approach that, despite Naess's actual words, points in a very different direction. Naess is an experienced philosopher and knows that complex questions emerge at every moment in his view of nature, and our obligations towards it and ourselves. Some ethicists, for example, have raised the question of whether it is really possible to practice the principle that all life forms have an equal right to flourish. Comparing the needs of, say, a mosquito and a human being throws the problems with this principle into particular relief.[22] Not only does there seem to be a disproportion between human needs and the life of the mosquito, but also it would be extremely difficult to see how any of the principles in the platform of Deep Ecology could give us any means to adjudicate between conflicting claims. To anyone who appreciates biodiversity, simply eradicating the mosquito for no good reason would certainly be wrong. But that draining mosquito-infested swamps and controlling mosquito-borne diseases have benefitted the human race, few would deny. At the most practical level, Deep Ecology seems of little use.

To his credit, Naess has tried to remedy this problem, not by a simple rule, but by a general attitude. He has conceded, perhaps under questioning by friend and foe alike, that human beings will have a special obligation toward their own species and that we all individually have a greater sense of responsibility toward those near to us. Most Deep Ecologists, however, speak and act as if the absolute equality of all beings could be translated directly into personal and social behavior—an extension to the realm of nature of the old utopian notions of egalitarianism. One astute political scientist has even claimed that the central problem with the Deep Ecology position is that, man or mosquito, no individual counts for much, only natural species.[23] Given the history of totalitarian political movements in this century based on enforced equality and group identification, there is no little danger in the submersion of individual persons in larger aggregations.

Part of Naess's view of the particular responsibilities of human beings involves a recognition of some differences in the worth of living things. He said once that if we are forced into the unfortunate position of having to choose to save a human baby at the cost of the last member of a species of endangered tigers, we should save the baby.[24] The philosophical bases for this position are not entirely clear in Naess's work, and the very need to ask the question raises alarms. But he grants himself a kind of escape clause from any of the inconvenient consequences of Ecosophy T by claiming that he is concerned less with developing an airtight environmental ethics than with stimulating an ecological sensibility. Given that sensibility, we may recognize a wide range of difficult choices that we must make, respecting vital human needs and obligations but always with a deep background sense of our commitment to the flourishing of all nature. In short, he winds up suggesting a cosmic muddling along.

More Humane Philosophies

Although Naess appears to reach that conclusion only reluctantly, his willingness to make even such a limited concession sets him apart from many philosophers of the environment today. Philosopher Curtis L. Hancock, for example, has described what has come to be the more typical position in academic philosophy. Species are put on absolutely equal levels, lest a pernicious "speciesism" lead us to think of ourselves as somehow more valuable than other beings. Hancock proposed the following scenario at a meeting of the American Philosophical Society: If upon entering a burning building you found a human baby and a caged squirrel inside and the situation allowed you to save only one of them and rush out, would you save the child? The answers were astonishing and point to several troublesome currents in the culture.

One professor said he would save the child, but only because "custom and law required it." When Hancock raised the issue of intrinsically greater worth, he got this convoluted answer:

True, the life of a squirrel seems readily expendable when balanced against a human life (although the infant makes it more controversial). But I could see a situation in which maybe twenty or thirty animals could outweigh in value the life of a human being. Perhaps if species were in the balance, a human being could also be sacrificed—say, to preserve the last strain of a bacterium or a virus? Also the quality of lives should be factored in. On balance the life of a retarded person, say, might have less value than, say, the lives of several healthy apes.[25]

When the lives of viruses and bacteria or even the higher animals can present so many challenges to the ancient notion of *Homo homini res sacra* ("Man is sacred to man"), and when the weak and infirm seem to be headed for redefinition as outside the human race, something is going deeply wrong with our views of nature.

Perhaps the central problem with Deep Ecology and related movements has been its relative emphasis on the principles of the platform and relative neglect of deeper commitments. Deep Ecology counsels a shift from anthropocentrism to ecocentrism. Given that ecocentrism appeals to diverse religious and philosophical "ecosophies" as well as giving rise to a diversity of practices, it is understandable that people living in a pluralistic society would try to find a common core, not of ultimate principles that are contested, but of penultimate principles around which many people might come together. But it is precisely because of this pragmatic move that the DEP needs constant scrutiny by people with ultimate principles, because human beings have never been content with merely penultimate premises. And for religious people in particular, the inevitable question presents itself of whether Deep Ecology is not in need of still deeper correction by a historically experienced theocentrism.

Basically, the issues revolve around three main points: antihumanism, theocracy, and a proper theological perspective.

Rebutting the Antihuman

In a kind of paradox, the generally greater material wealth of the developed world is gradually drawing many different kinds of people into a sense of a responsibility for the smallest living things, a sense that is absent from poorer societies. This sensitivity manifests itself even in some surprising quarters. For example, the conservative British writer Paul Johnson rejects, as might be expected, the more radical, antidevelopment strains in much of modern environmentalism. He warns against the neopagan side of environmentalism and mistrusts Greenpeace, Friends of the Earth, the Club of Rome, and other self-styled defenders of the planet as substituting nature worship for real religion.[26] One of the results of this new form of faith is a shift from a previous religiously inspired balance between man and nature to the current view of human beings as a kind of natural scourge: "The early environmentalists, being mostly enthusiastic Christians, were never anti-human." Without the theological sheet anchor, says Johnson, environmentalism has a tendency to "slip into extremism and attract fanatics."

Yet for all these qualifications, Johnson himself feels that a shift in consciousness is occurring, not in opposition to, but as a result of, technology: "Our technology is now such that we can produce endless varieties of nourishing and delicious foods without resorting to animal flesh. . . . I believe we shall gradually come to regard the eating of animals as no more acceptable than cannibalism—and no more necessary, either."[27] Not only has Johnson given up hunting as a result of this growing realization, but he also argues, "An ordinary house-fly, closely inspected, is a miracle of contrivance. It now seems to me that to swat such a remarkable being, except under the clearest necessity, is an outrage against nature." Johnson is not at all a unique case among more conservative religious writers on the environment. He shows, if it needs showing, that anyone who examines many of the pro-capitalist, pro-development thinkers on this subject will find, not simply a blanket defense of industrialism and a callous attitude toward nature, but a sensitivity to some of the same concerns that Deep Ecologists think exclusively their own preserve, though with a different program for how to handle problems. A Johnson might argue, for example, that a less affluent society will lead us back to the very necessity to treat nature roughly that the Deep Ecologists deplore.

But even beyond the pragmatic considerations, a better balanced effort to relate the deep and the shallow must really attempt to engage ultimate commitments, whether Jewish, Christian, Muslim, Buddhist, or philosophical. Almost every person on earth has such commitments, sometimes in clear form, sometimes in relatively confused fashion. As the Apron Diagram itself indicates, there should be a flux and reflux between specific things that we believe need to be done and the kind of illumination our ultimate convictions supply. The besetting sin of Deep Ecology is to see the traffic as going mostly in one direction. Despite all that has been written and said about drawing on "fundamental premises and ecosophies," Deep Ecology seems to select quite capriciously from the deeper strata on the basis of the kinds of environmental action closest to the hearts of Deep Ecologists. There is very little in the way of opening up to traditions that have faced many different situations and civilizations and have proved their staying power.

Naess's appeals to Buddhism and Spinoza, for example, show relative neglect for the portions of that kind of thought that see something unique in human beings, even if we are not in the final analysis to be viewed dualistically as different from nature. That narrowness accounts for the cramped space allowed human life in Deep Ecology. Human beings leaping between rocks to avoid ecological damage is just the beginning of the problem. Saint Paul counseled human beings to "work out your salvation in fear and trembling." Perhaps an analogous circumspection should govern our concerns about nature. But it would be difficult to believe that nature or God put us on this planet to exhibit a kind of cowering timidity in the face of what we might do to it.

When we really go back to the deeper levels of commitment, we may find that—as Naess occasionally concedes—human beings have a special, architectonic place in nature's order. The willingness of some Deep Ecologists to resort to virtual war on behalf of nature bears witness to that truth in perverse fashion. Most religions do not believe in a simple pacifism. In just-war theory, for instance, criteria have been developed to help us to know when it is morally proper to use force for the purposes of justice and self-defense. Carefully articulated systems like that have grown up around religious doctrine precisely because such issues are too serious to leave to the whims of vigilantes who believe that they may take matters into their own hands on their own authority. Yet there is a time when the use of force in a good cause becomes justified. Deep Ecologists of certain stripes may, then, be right about that in theory. But their practice is far too casual, precisely because they do not respect the kind of articulated moral reasoning found in all civilizations. At the same moment that they proclaim nonviolence, Devall and Sessions, for example, have allowed "the decommisioning of a power generator or bulldozer," just as long as they are "spontaneous acts.[28] This incoherence may be more than a momentary lapse. Though some Deep Ecologists display great powers of rational analysis, others like Devall show a fear of the powers of reason: "deep ecology is best expressed, not explained."[29] It is precisely to prevent such "spontaneous" self-righteousness that religious and philosophical ethical traditions have been created historically.

Deep Theocracy?

Issues like these may lead us to believe that a great deal of what has constituted the best parts of civilization is simply being brushed aside by Deep Ecology. It has taken enormous historical experience for the human race to put in place the principles that restrain our propensity to give vent to anger and self-righteousness. Tolerance of other views, for example, which Naess often speaks of in principle, in practice seems to give way to the urgency of implementing the eight principles of the Deep Ecology Platform, despite legitimate opposing points of view. As Martin W. Lewis remarks in his book *Green Delusions,* "While celebrating diversity and toleration, deep ecologists dismiss virtually the entire heritage of Western thought as not only bankrupt but as actively leading us to ecological destruction."[30] Civilizations may become corrupt or unbalanced, but so may efforts to reform civilizations. Particularly when living human beings, with the sacred rights of their consciences and intentions, are ruled out of order or ridden over roughshod, we may detect cultural criticism entering into a kind of ecological theocracy.

This reflection leads to another. Probably the most attractive feature of the Deep Ecology program is its claim to value inputs from multiple sources and to wish to promote pluralism. But when we look at the voices they are actually willing to listen to, the range is quite small and the views they are willing to heed quite narrow. As we saw in the previous chapter, similar scientific views may lead to strong cases for diametrically opposed policies. Deep religious commitments can lead to equally unpredictable environmental stances.

Naess seems to be practicing a strategy that would enlist a variety of religious denominations in a crusade that already knows its goal. There is little room, for example, for the kind of religious voice that might say that our God-given capacities for innovation and technological advance may exist precisely because of the challenges the Creator knew we would face, early on in the struggle with nature, later in thinking about how to keep our numbers from causing undue harm to nature.

It is only because Naess has not really engaged in dialogue with the kinds of believers who think that the right sorts of development may help the environmental situation, for instance, that he can make as simple-minded argument as the following: "One per cent increase in 'Gross National Product' today inflicts far greater destruction of nature than one per cent 10 or 20 years ago because it is one per cent of a far larger product. And the old rough equivalency of GNP with 'Gross National Pollution' still holds. And the efforts to increase GNP create more formidable pressures against environmental policies every year."[31] Even in mere statistical terms, growth in developed societies is more and more occurring with lower per capita consumption of fuels and raw materials. The wooden view of economics here needs opening up to other points of view.

But it may be all but impossible for Deep Ecology to make that move because the whole thrust of the Apron Diagram is to filter theological principles through an environmental program rather than the reverse. Every conceivable form of religion is made to pass through the narrow waist of the Eight Points, points that themselves need to be opened wider to religious and philosophical speculation. For all its claims of pluralism, Deep Ecology runs the risk, in the final analysis, of being a kind of theocracy. Like earlier theocrats, Deep Ecologists banish to the outer darkness those who do not fit into the Deep Ecology principles. It would be difficult to see how any committed Deep Ecologist, for example, could countenance the trade-offs (temporary, we hope) of harm to ecosystems for preservation of human life that will inevitably face us in the near future. The emphasis on nature is great, the regard for the human small. The god of Deep Ecology is a cruel god, who is not made any more likable because he presumably feels greater kindness towards nature.

Deep, Deeper, and Deepest

If the notion of depth has any meaning, it is that the ultimate reality of the universe will tell us some things we do not expect to hear. All over the world, that voice has advocated doing justice, showing mercy, feeding the hungry, tending the sick, liberating captives, and a long list of other compassionate acts. In some religious systems, Buddhism and Christianity among them, compassion for the suffering of animals appears. Buddhists seek to eliminate all such suffering. But most Christians have decided that there are uses to which animals may be put for human food and well-being. The 1994 *Catechism of the Catholic Church,* for example, strikes a balance based on deep theological beliefs. On the one hand, it says, "It is contrary to human dignity to cause animals to suffer or die needlessly."[32] The reference to human dignity is important. It reminds us that the way human beings treat animals should be a reflection of their own status as beings called upon to do the divine will. In his encyclical *The Hundredth Year,* Pope John Paul II extends this principle by linking disregard for "the natural habitats of the various animal species threatened with extinction" to a failure to "safeguard the moral condition for an authentic 'human ecology.'" Again the human and the nonhuman are intertwined, and both are viewed through an ultimately theological lens.

But in neither case does this simply require us to keep hands off animals or other parts of the creation. Arne Naess counsels the absolute minimum of death and interference. The Catholic Church takes a much more unapologetic view. The *Catechism* argues that within reasonable limits, medical experimentation on animals and other uses of animals are morally justifiable because it is not just to give to animals the same degree of respect proper to human beings.[33] Buddhism aside, that seems to reflect the basic wisdom of truly deep religious traditions, neither failing in care for nature nor ignoring human importance in favor of a false and ultimately sentimental identification of man with the rest of nature.

It is not necessary to take the specifically Catholic view to recognize that various religious positions deeper than those of Deep Ecology's platform may be equally or even more reasonable, practical, and finally just. A real engagement with the deepest human intuitions and commitments would present us with a much richer picture and program than the one Deep Ecology has been laboring to construct. We will not leap between rocks like goats, because we are not goats but human beings with our own way of treading the earth. Without question, the religious traditions need to be opened up to the new situation in which we find ourselves. But at least as much as innovation, we may need the steadying influence of various religious traditions *in their entirety* to keep us from overlooking parts of wisdom that have been accumulated patiently over the ages.

Deep Ecology shows a breathtaking confidence in its own ability to criticize the entire Western story without having as yet shown itself capable of the kinds of prudent arrangements and delicate discriminations that the West has at times produced. In some ways it is quite easy to criticize the past, because previous philosophies and theologies are bound up with societies that were quite imperfect. We know the bad of the past; we tend to underestimate the good. A movement like Deep Ecology can make sweeping claims without having had to face the consequences of those claims over long time periods. The ancient Greek philosopher Aristotle once said, "Little errors in the beginning mean great errors in the end." A movement that values the slow, geologic processes that have resulted in our world must constantly remind itself of the hidden wisdom also embedded in human thought and the dangers that may lurk in assuming that what seems certain, necessary, and indisputable at this moment is the sum and substance of true wisdom.

5

The Gospel
According to Matthew

Religious reflection on the environment goes on at several levels of sophistication, from the most abstruse theological and scientific analysis to the numerous popular handbooks currently available about how to care for God's world. All the major religious bodies have issued statements on the environment, usually quite middle-of-the road on their face, reminding adherents of their obligations as the world's stewards.[1] Occasionally, one of these official statements will make sweeping claims, such as the National Council of Churches' (NCC) early characterization of global warming as a "spiritual crisis." In 1998, a coalition led by the NCC even petitioned President Clinton to seek U.S. ratification of the 1997 Kyoto Protocol for addressing climate change. But by and large religious denominations have taken a measured and, perhaps, even overly passive approach to environmental challenges. Typically, they neither educate people about the theology of creation nor motivate them to act on specific problems. At the parish or popular level, then, there is some initiative for environmental action, but not very much and only of the most basic kind.

By far the most powerful impact on religious thinking about the environment at popular levels has come from activist groups and individuals. One of the most prominent and influential of all of those figures is Matthew Fox. Fox was for many years the most visible Catholic theologian writing on creation and, therefore, on environmental issues as well. He might still be, if conflicts with his superiors had not led in the mid-1990s to his dismissal from the Dominican order and his subsequent departure from the Roman Catholic priesthood and the Catholic Church. Though Fox now describes himself as "postdenominational," he also — only God and the local bishop know how — serves as an Episcopal priest in the San Francisco Bay area, where his earlier Institute in Culture and Creation Spirituality flourished at the Holy Names College in Oakland and seems now to have grown into the University of Creation Spirituality. About half the students at these institutions — despite a faculty that includes self-proclaimed witches, Yoruba voodoo priestesses, Celtic and Druid practitioners, and other spiritual exotics — are priests, nuns, and other religious. Before the break with the Catholic Church, Fox had received support from Catholic institutions for many years and exerted enormous influence on women's orders and directors of religious education, two groups that strongly shape church programs. And he has frequently led conferences, retreats, spirituality workshops, and alternative, earth-spirit liturgies around the country and in international forums. Though the liberal *National Catholic Reporter* (NCR), usually sympathetic to his battles with church authorities, once described Fox's views as "a combination [of] Doonesbury cartoon theology and Shirley MacLaine spirituality,"[2] through his energetic proselytizing Fox has probably shaped American religious thought and practice on the environment at the popular level more than any other single figure.

Yet whatever else might be said about Fox's views, it is necessary to recognize from the outset that he deserved his visibility. The NCR characterization was partly unfair. Fox on spirituality often sounds like Shirley MacLaine, but he actually knows something about the Western and Eastern spiritual traditions. In addition, no other modern religious writer has attempted anything like his comprehensive, energetic, and sometimes insightful recovery of traditional elements that provide us with a renewed way of both thinking and feeling about nature, including human nature.

The Saintly and the Trivial

But Fox obscured this more substantial side by the titles he gave some of his earlier works such as *Whee! We, Wee All the Way Home: A*

Guide to Sensual, Prophetic Spirituality and *On Becoming a Musical, Mystical Bear: Spirituality American Style.* In the Gospel according to Matthew Fox, the playful *puer* (aka the inner child) must be restored to a theology and spirituality too long dominated by the solemn *senex* (wise elder). It is only one indication of Fox's weakness for anything that presents itself to him as mystical and celebratory that he believes American churches and society at the end of the twentieth century are dominated by a kind of patriarchal senility, something not self-evident to most cultural observers, especially around Fox's San Francisco Bay area.

Even more significantly, however, Fox has been the frequent leader of "liturgies" that draw on the most bizarre impulses in modern American New Age movements coupled with idiosyncrasies all his own. Like many religious environmentalists, Fox is fearful of disturbing what he believes to be the natural, God-given balance of nature. Native peoples who have similar beliefs, he asserts, create rituals closely attuned to the regularities of nature, which they often think need to be performed so that nature and man do not get out of sync. Fox admires such peoples and their practices, but when it comes to rituals that we might want to perform today, he innovates with no regard for tradition or sensibilities — biblical or indigenous — that might make ritual more than a kind of free-floating symbol for whatever he happens to find of interest at the moment. Real ritual carries a good deal more weight in any religious system than that.

For example, he is fond of advocating liturgical dancing in circles, "Sarah's circles" as opposed to rituals organized according to what he calls the hierarchical ups and downs of "Jacob's ladder." Women religious, it appears, often respond to these liturgies, especially when it is coupled with goddess talk and Wicca invocations of women and nature. Fox has little use for existing churches, believing alternately that it would be better for us either to get out to the great outdoors or to retire to caves to reacquaint ourselves with nature mysticism. The cave liturgies, he says, would return us to the kinds of practices early man carried out, as we may read them in cave paintings, which, Fox assures us, are identical with the most advanced thinking about mysticism and harmony with nature. At other times, he suggests liturgies to celebrate the parts of the body — a liver Mass, say, or others devoted to sexual organs, heart, or lungs. There is, of course, a weak symbolic value in all of these made-up liturgies. But it is precisely their spur-of-the-moment, superficial nature, unanchored in any historic religious tradition (except for alleged nature religions like Wicca), that makes them of little weight religiously let alone environmentally, to anyone outside of Fox's circles. That inside those circles, however, they have been for many people a moving experience is an indication of a deep weakness in certain sectors of mainstream Christianity and Judaism.

Fox argues for the validity of this California-style innovation on the grounds that historically the West has undergone several hundred years of overemphasizing the Fall/Redemption component in theology (another notion that does not seem dominant in our current cultural context). He posits an Original Blessing anterior to and more basic than Original Sin. The Original Blessing in the Creation demands the development of his central principle, Creation Spirituality, which he has presented in several books — *Original Blessing: A Primer in Creation Spirituality; The Coming of the Cosmic Christ; Creation Spirituality: Liberating Gifts for the Peoples of the Earth;* and *Sheer Joy: Conversations with Thomas Aquinas on Creation Spirituality.*

A History of Spirituality

In broad outline, Creation Spirituality's reading of history goes something like this. Prior to the seventeenth century, people in most cultures maintained a mystical sense of their oneness with creation that was expressed in their cosmologies and rituals. The rise of modern science destroyed that old wisdom in the West, and as a result, the cosmological dimension of Christianity simply dropped out of sight. Fox picked up some of this historiography from the great twentieth-century theologian M. D. Chenu, with whom he studied in Paris as a young man. Chenu thought that the Scholastic philosophers who had succeeded Thomas Aquinas in the sixteenth and seventeenth centuries had betrayed the master by a failure of nerve. Instead of seeking to incorporate the truths of reason discovered by modern science into the old wisdom, they had simply abandoned the task, giving Christianity an inward-looking overemphasis on sin and salvation. According to Fox, most Christian churches today continue to make the same mistake.[3]

Obviously, judging the truth of such complicated matters about historical periods presents formidable difficulties. Among them, it would be difficult to take to task the theologians criticized when the best reason of the time seemed to have done away with cosmology in the old sense. Newton himself was a believer who thought his discoveries revealed the real workings of the creator, as opposed to what some people had imagined the creator had done. Philosophers since the seventeenth century — Descartes, Leibniz, and Kant, among others — may be read in some ways as attempting to preserve the old human values in the absence of any coherent cosmology. Only in this century, with the discovery of the expanding universe and the initial singularity that gave rise to the world, has it become possible to put science and religion into rough cosmological harmony again. The theologians whom Fox and Chenu deplore may have simply been doing the best job they could under unfavorable circumstances.

But if those theologians represent one, perhaps undesired, extreme, Matthew Fox quite definitely represents another. At times, he seems determined to make up for lost centuries of cosmology with a vengeance. Like Thomas Berry, Fox has worked with the mathematical physicist Brian Swimme, and he has also looked into some new biological theories in dialogue with scientists such as Rupert Sheldrake, with whom he has also written a book.[4] His understanding of this science does not go very deep, and he often misunderstands or misuses scientific terms as if they were an exact confirmation of his theories instead of suggestive analogies to be approached with some caution. But for Fox, the new cosmology tells us that we all began in the same sudden time-space event fifteen billion years ago, and that is enough to authorize us to believe in a common creator and to seek his presence in contemporary mysticisms.

Mystical Pan-en-theism

Mysticism, ancient and modern, plays a large role in Fox's system. In addition to his own work on Original Blessing and Creation Spirituality, he has compiled, or inspired others to compile or retranslate, anthologies of works by figures such as Hildegard of Bingen, Meister Eckhart, Dante, Juliana of Norwich, Mechtild of Magdeburg, Nicholas of Cusa, and even Thomas Aquinas. Fox finds in all of them two main features: They all had a premodern view of the cosmos that saw the world as sacred, and they viewed man and nature as linked by microcosmos/macrocosmos correspondences. Thus, they give us a way to reconnect the older Christian tradition with modern cosmology, jumping over the bad old Enlightenment days of Descartes, Newton, and Kant, for whom the world was a mere mechanism and God something outside it and us. Fox redefines the earlier medieval figures not as theists but as panentheists.

Theism, for him, suggests a wholly transcendent God who is at best weakly in touch with the universe and human history. In Fox's view, the spiritual life, in that kind of theistic religion, consists in a Platonic attempt to leave the world of matter behind for purely spiritual contemplation. The universe is thus simply regarded as a fallen realm, a vale of tears that we should shun and escape as soon as possible. Fox states baldly that "if the only option I was given by which to envision creation's relationship to divinity was theism . . . I would be an atheist too."[5] Yet this is hardly the meaning of theism for any of the major figures in the Christian tradition of the West. Most of them would have regarded such a view as a concealed Gnosticism. And Fox, who displays no small theological and historical knowledge, must know as much.

The kind of theism he describes here and in many other places in his work belongs more to the Enlightenment departure from Christian thinking in certain quarters. Almost all Christians continued to believe that God designed the world and exerted providential guidance over it; they prayed to him as active in history, and worked for greater justice and material well-being in this world precisely because of their Christian beliefs. The early Enlightenment figures did limit nature to something that we now see, even on scientific grounds, as highly reductionist. But very few Christians thought of God as the Supreme Watchmaker who simply wound up the universe and then let it run blindly, granting no value or meaning to human life other than its mechanical order.

Indeed, in reaction to that exaggerated materialism, several philosophers—among them Spinoza, Leibniz, Fichte, Schelling, and Hegel—elaborated systems of idealism in which God became closely identified with the universe. That philosophical current, usually called pantheism, creates any number of difficulties for a biblical view of God, not least that if this universe we see before us is part of God, all the evils within it — murders, rapes, famines, wars — are essentially part of God as well. The traditional view was that God had created a world of secondary causes and watched over it, but that he permitted human misdeeds because he wished us to have free will, which we might use or misuse. He also permitted natural evils, whether as a punishment for sin, as a consequence thereof, or for some obscure reasons, as we saw in our earlier discussion of Augustine. In any event, theism is, among other things, the Bible's way of drawing a distinction between God and natural evils.

Panentheism, Fox's preferred term, was proposed by Karl C. F. Krause in the early. nineteenth century, probably in reaction to Kant. Clearly, it tries to have things both ways. In Greek, panentheism simply means "everything exists in God," not that God is everything, as in pantheism. At first sight, this seems to solve two modern problems. It would restore to nature a sacred status, since all creatures including human beings are in God and God is in all things. But it does not make God identical with every part of creation, since nature is an articulated structure of good and bad. Yet panentheism does not entirely avoid the problems associated with pantheism and may be more of a semantic than a real solution. If the earth is "in" God and God "in" the earth, earthquakes, tidal waves, cancers, and AIDS are closely tied to God's nature. It may be possible to introduce nuances to save the whole conception. But unless we wish to abandon the biblical view of God as a purely and supremely good being, there has to be more distance introduced between him and his Creation than Fox allows in his version of panentheism.[6]

The Cosmic Christ

One of the reasons Fox is not wary of the identification of God and nature is that, like many in the environmental movement, he sees nature as far more benevolent than, in fact, it is in most human beings' experience. Take, for instance, this description, in which Fox argues that we should cease regarding Christ as the Light of the world and start thinking of him as Mother Earth, who is crucified but rises daily:

Like Jesus at Golgotha, she is innocent of any crime. She has blessed us for four and one-half billion years by providing waters; separating continents; establishing just the right amounts of oxygen, hydrogen, and ozone in our atmosphere for us; birthing flowers, plants, animals, fishes, birds to delight us and bless us with their gifts and their work of making air and soil healthy and welcoming to us. In short, earth loved us — and still does — even though we crucify her daily.[7]

This is a beautiful expression of gratitude towards nature and a powerful reminder of how we depend on the interrelationships of everything in the cosmos for our existence.

Yet it leaves out at least as much as it includes. If we were not already aware of the struggle human beings have had with wild animals, disease, climate, famine, and other human beings, we would assume that the world Fox is describing was literally arranged for human comfort — a view that might be thought an immature, "fundamentalist" reading of the Bible. No native tribe, the very people Fox idealizes everywhere in his work, would send its young men and women out into the world equipped only with this sort of vision. All serious religions combine gratitude toward creation with a healthy fear and respect for its dangers. Native peoples teach survival skills and harden their young to endure precisely because nature, even in the welcoming and specially suited environment of this earth, challenges us in ways that Fox rarely seems to recognize. Even the Western mystics who form the core of his system often show more ambivalent attitudes towards the earth than Fox acknowledges.

But Fox has something more in mind in his mystical panentheism than a mere correction of an abstruse question about God and nature. The main element that unites all the mystical figures he admires is that they regarded the truths of religion not only as external theological propositions to be understood but as spiritual principles to be lived and felt experientially. The problem with our churches, including the problem of attracting and retaining the young, is that they repeat formulas without the cosmic connection that vivifies them. Recovering that cosmic mystical sense, he believes, would open the door wide for us to refashion prayer, liturgy, art, politics, economics, agriculture, industry, religious bodies, and our sense of ourselves in the world. For Fox, mysticism is the true reality of which "religion" is merely a report. Mysticism heals sin, mind-body dualism, denominational division, and man/nature conflicts.

The Watering Hole

Creation Spirituality, however, does not proceed, as do older theories of cosmology and natural law, from a living understanding of creation to a deeply felt set of commands about what constitutes God's intention for us now. Instead, Creation Spirituality seems to emerge at the intersection of feelings already poised to leap forth within diverse groups:

> As a movement, creation spirituality becomes an amazing gathering place, a kind of watering hole for persons whose passion has been touched by the issues of our day — deep ecologists, ecumenists, artists, native peoples, justice activists, feminists, male liberationists, gay and lesbian peoples, animal liberationists, scientists seeking to reconnect science and wisdom, people of prophetic faith traditions — all these groups find in the creation spirituality movement a common language and a common ground on which to stand.[8]

This list of creatures around the Creation Spirituality watering hole forms a kind of honor roll of countercultural causes and groups, though it is difficult to see how they fit easily together except as a poetic counterforce to everything Fox deplores as the mainstream of the modern West. Native peoples are often used as a political ally for current anti-Western movements. But respect for native practices and beliefs should lead us to recognize just how distant they are from current "men's movements," "gay and lesbian peoples," "feminists," and "animal liberationists." It is only by focusing on a putative mystical oneness and ignoring actualities that, say, specific groups of native peoples — with their tribal conflicts, hunter-warrior ethos, and cosmological ball games — do not more easily line up with the U.S. Marines, Rambo, and the National Football League.

The vaguely chic, New Age dimension of Fox's project is particularly unfortunate because, whatever entrée it may offer to rather dubious elements in contemporary culture, it obscures much that is valuable, even invaluable, in his analysis. When he remarks, for example, that "there can be no anthropology without cosmology,"[9] he is stating a truth that modern people living in developed societies, primarily in cities, need to know. Christianity and Judaism have often been criticized for their "anthropocentrism" in which the earth itself was thought of as the center of the universe. Fox rightly reminds us that it is precisely the noncosmological concentration on human affairs divorced from God and his creation since the end of the Middle Ages that has created a far more radical and incoherent anthropocentrism than existed in premodern societies, whether Western or non-Western.

The Four Paths

Spirituality of a certain kind, something more than the revised history of the universe as explained by modern scientific cosmology, is necessary to understand where human beings fit in. Fox identifies Four Paths of Creation Spirituality: (1) the *Via Positiva* (Positive Way) registers our awe at the existence, complexity, and beauty of the universe; (2) the *Via Negativa* (Negative Way) recognizes our emergence from cosmic darkness and the shattering experience when all we trust is shaken, as a prelude to greater knowledge and compassion; (3) the *Via Creativa* (Creative Way) is the way in which we most express our creation as the *imago Dei* through our own creativity; and (4) the *Via Transformativa* (Transformative Way) represents our efforts to reestablish justice and reach homeostasis, or the equilibrium of all physical, moral, and spiritual forces in the universe. Especially after the Negative Way, the *Via Transformativa* also leads to universal compassion.

As a general statement of spiritual life, the Four Paths take in a lot of what makes human life human, but they are also significant for what they do not include. As we have earlier demonstrated, homeostasis, or equilibrium, does not really reflect the cosmic process, except locally and briefly. Fox believes the opposite, though on what spiritual or religious grounds is not entirely clear: "Homeostasis, the scientific word for justice or equilibrium, is now recognized as a basic principle of the universe."[10] Other than certain rough stabilities in earth's biosphere, however, the very dynamic, creative, destructive, reconstructive universe that Fox admires can hardly be described as exhibiting homeostasis. As a result, the Creative Way and the Transformative Way, as Fox has formulated them, do not give us very much guidance about how to be "creative" and "transformative" except within a presumed steady state, if such a thing is possible. Fox himself halfheartedly recognizes as much when, after the poetic evocation of creative satisfactions, he vaguely speculates, "Being so young a species on this planet, with immense powers of creativity, we need ways that help us guide that creative energy in directions that allow our passion to mature into compassion."[11] But what are those ways if our only aim is homeostasis?

An even more glaring omission from this fourfold spirituality is any mention of how we are to provide for our material needs. Because Fox's Negative Way is limited to contemplating darkness and loss and his Creative and Transformative Ways have mostly to do with art, ritual, and social justice, he never directly confronts the everyday business of shielding ourselves from natural threats and laboring to increase natural goods. Indeed, even in *The Reinvention of Work,* a late book apparently intended to remedy the lack of attention to human labor, Fox is long on making work meaningful and fun, short on recognizing how we earn our bread by the sweat of our brows.[12]

Like many people who have not thought much about creativity in economic sectors, Fox has a rather static notion of how economies grow and what constitutes resources. "At whose expense will the economy grow? How can we have infinite growth on a finite planet without someone or something having to pay a dear price? And isn't that exactly what industrial societies have subjected the planet to—an infinite plundering of limited resources of fossil fuels, forests, water, air, plants, animals, people?"[13] The long answer to this last question is: no. To begin with, Fox presumes that wealth is not created, merely extracted and divided. But thanks in large part to the rise of the computer industry and other technological innovations, growth in the future may mean enjoying more material wealth with less impact on the planet. Similarly, to speak of "plundering" resources, as if oil or gas reserves or metal ores belong by right to "the earth" and should not be put to useful purposes, is an anthropomorphizing that will not stand much scrutiny. Economies may grow without expense to anyone or anything.

Fox also reads contemporary politics through this homoeostatic cosmological lens. The poor are poor in the Third World because the rich are rich in the First World. The debt crisis in Latin America, for instance, is a further instance of racism, imperialism, and the European exploitation begun in the age of discovery. We should cut the military budget $100 billion, which will enable us to fund social programs and provide jobs. It may also stop the shameful U.S. interventions everywhere.[14] No contemporary economist or political scientist of any standing would be likely to read the global political and economic environment the way Fox's mysticism leads him to.

Mystical Environmentalism

In general, Fox has a very weak grasp of current environmental issues. He picks up large numbers from various sources that are supposed to make us worried about our impact on nature. But the numbers often seem to be drawn out of a hat and pushed to absurd extremes. At one point, he accepts extreme-end estimates of species loss, and argues, "At this rate humankind will eliminate ten percent of the remaining species (one million of the remaining earth creatures) in the next ten years," that is, during the 1980s, a time in which no scientist has claimed anything like that number of extinctions. He immediately follows with the claim that "if current rates of destruction continue, within the next one hundred years there will be no living species left on this planet — including humankind, since we are totally interdependent with all these other creatures."[15] During the Cold War, antinuclear activists tried to convince us of the folly of thermonuclear weapons by asking us to imagine a world where cockroaches alone survived. For Fox, human environmental impacts seem capable of doing what even nuclear bombs, killer-rock comet strikes, and vast geological shifts were unable to do: kill the roaches.

According to people who have attended Fox's lectures, he has also variously claimed that within thirty years "Iowa will be a desert"; within twenty-five years the United States will be importing food; within fifteen years "one-third of all the world's forests will be gone"; within fifty years "in the United States, not one tree will be standing."[16] It would take no small effort to find anyone in the secular literature who would be willing to support a single one of these claims. But that may be because for all his claim of concern about the earth and his desire to encourage creation spiritualities that will root us once again in nature, Fox shows little concrete interest in the kinds of issues that constitute the environmental question.

If cosmology or cosmological rituals or cosmological mysticism were enough, the cosmologically rich cultures Fox admires would have lived either in the Garden of Eden or in the New Jerusalem. But whether we look at pre-Columbian native America or pre-Enlightenment Europe, aboriginal Australia or Celtic "goddess" cultures, we human beings are a pretty sorry lot. Wars and rumors of wars, sexism, environmental degradation, slavery, deadened personal perception, and authoritarianism coexisted quite nicely with the mystical visions Fox thinks a panacea. Sometimes the mystical visions were interwoven with, say, tribal or religious warfare. Original Blessing may indeed have preceded Original Sin, and we should all be seeking the reign of the former over the latter — at the end times. In the meantime, Fox's visions are a new utopianism that is not simply wrongheaded about how we will live our lives but a deep threat to the decencies we have managed to wrest from the butcher block of history.

Fox has a large capacity to believe passionately in idealized, far-distant exemplars. On archaic societies, for example, he claims: "As we know, in that five-thousand-year period in Europe when they worshiped the mother goddess, we have found no evidence of any military artefacts."[17] As we shall see in the following chapter, we know no such thing. Prehistory means a period before history, where human cultures are even more difficult to interpret than the ones for which we have written records. Fox posits a whole human Eden that, given our primate forebears' quite aggressive and male-dominated behavior, quite likely never existed.

It is easy to say that primitive peoples with cosmological beliefs and sentiments point the way for us to live in harmony with nature. It is much more difficult to specify exactly what that means. If we look, for instance, at the vast Midwestern prairies in the United States, which Fox and many other people believe are one of the natural gifts that we may be threatening by unwise agricultural practice, we discover something quite curious. A recent survey of the environmental literature concludes that the prairies "would have mostly disappeared if it had not been for the nearly annual burning of these

grasslands by the North American Indians."[18] Grasslands have a natural tendency to undergo a progression into woodlands. So here we
have an instance of native people deliberately restraining nature from
taking its "natural" course.

Native Americans had reasons to prefer the open spaces to forests.
Perhaps it made hunting animals easier. Perhaps it made it easier to
spot enemy advances. But it is curious that we assume their practices
were benign and natural, ours detrimental and artificial. If a large
multinational agribusiness were today to do what not only North
American Indians but native peoples around the world have done —
aggressively burn forest edges to increase clearings over hundreds of
miles — we would regard it as an outrage. Our anger would probably
be justified when it comes to the tropical rain forests, where many
undiscovered species are thought to reside. But that very case reveals
how much different — and better — even our secularized scientific
approach to the environment may be at times than a sentimentalized
and facile belief that native ways, rooted in cosmology, can save the
planet. They cannot. And several of them, from slash-and-burn agriculture to traditional ways of farming, may actually place greater burdens on the biosphere than their modern counterparts.

All Are One?

But goddess eras, Native Americans, aboriginal peoples, and the
various religions of the world must all be One Good Thing at the
Creation Spirituality watering hole, though diverse, of course, too,
because we appreciate diversity even though we are all the same. Fox
carries a common contemporary belief — that all religions are really
just different paths to the same end — to an uncommon pitch. Even
on the surface, it might appear that this supposed melding of all traditions lacks a real foundation. A Southern Baptist, for example, may
imagine heaven to be a realm of clouds with angels and harps, or
something like a revival meeting in church, or maybe a joyful picnic
on a golden summer day. But he or she will look upon, say, the Zen
Buddhist desire to pass beyond all duality as a different goal, and perhaps as a snare of the Evil One. The Zen Buddhist may feel the same
about the Baptist's desire for personal immortality and eternal happiness. Only if you define at the outset that all religious positions are

"really" about a specific kind of mysticism—something ordinary believers and authorities in those traditions may dispute—can you have
the kind of religious oneness that Fox assumes.

The same is true of the environmental effects of various religions.
Fox's own vision of a dynamic spirituality coming forth from the
great Spirit that blows where it will can lead in many directions. For
some people, it will lead to a very quiet contemplative life, a perfectly
legitimate response for them. For others, it may lead to active labors
on behalf of others. Fox admits this latter possibility, but for him it
mostly means "social justice" activism. The divinely willed need for
us to earn our daily bread barely puts in an appearance in Fox. For
him, listening to nature means contemplating beautiful forms. But
listening to nature may also be listening for penicillin, for ways to restore eyesight, for new species of plants for food and of animals waiting to come into existence, for mechanical inventions that relieve us
and nature of some of our burdens.

All religions are not equal in this respect. There is a reason why
science never flourished among Buddhists, for instance. Like the reasons for the emergence of science in the West, those reasons are multiple. But it would probably not be too far a reach to say that a religion that believes in the world as God's creation, like the religions of
the Book, will take an interest in the physical world. And historically
speaking, at different periods, Jews, Christians, and Muslims led the
development of science. A religion that believes the world is *maya,* or
illusion, will not pay as much attention to the world. Scientific development, of course, is no measure of the rightness or wrongness of a
given faith. But it is worth emphasizing—in the face of all Fox's appeals to mysticism—that spirituality of the kind he admires may lead
to human squalor and natural degradation as easily as to some imagined harmony with a benign nature.

A Mature Sixties?

But Fox's broad popular appeal has always been only tangentially
related to the debatable ideas and judgments to be found in his
books. He has had his greatest influence in advocating a certain spiritual lifestyle, and that advocacy has burned like wildfire through several religious communities. It is no secret that mainline Christian

and Jewish communities have been shaken by the social revolution of the past three decades. There are two ways to read that revolution: either as a rejection by materialistic cultures of the age-old restraints of religion and philosophy, or as a liberation of human beings, finally entered into their maturity, from oppressive social, ethical, and religious rules in order to embrace the divine spark within. In the latter view, we are still much oppressed by dead forms all around us and in our churches. The pressing need is to throw off lifeless tradition for living self-liberation in mystical ecstasy along with nature.

Fox invokes images of this kind so often that a person unfamiliar with the Bay Area or various sectors of American religion today might get the impression that we are suffering from widespread repression to which Fox speaks as a prophet. In fact, we are suffering from widespread self-indulgence and—somewhat outside his intention—Fox plays precisely into a current mood. For all its rejection of past follies, the current age may be exactly what he describes it as: "I expect the nineties to be regarded in time as a mature sixties."[19] What "mature" might mean in this context is a matter of dispute. Certainly Fox countenances no illegal drug use. He is largely traditional in his dismissal of sexual promiscuity. And his firsthand acquaintance with some of the great mystical writers of the world saves him from the worst self-congratulatory excesses of the 1960s gurus. But that said, he embraces virtually every theory that did not self-destruct thirty years ago.

This leads him to misconstrue several problems. For example, he thinks that our violence towards nature and one another is the product of a bored culture. But as native peoples throughout history might tell him, violence is often the product of threats, and threats among tribes rise up even when their cosmologies are congenial to Matthew Fox. The error here is a logical one that more careful thought and less mere enthusiasm would have helped him avoid. While wrong cosmologies can induce wrong behavior, right cosmologies need patient elaboration through philosophical categories like wisdom and prudence, and political and economic structures that can only approximate the good and keep evil in check. The very emphasis on the universe as a totality in Fox leads him into a kind of spaced-out feel-good environmentalism.

The weakest part of Creation Spirituality is its ethics. It tends to decry current beliefs as excessively moralistic on the one hand, or insufficiently social on the other. Scientific materialism is a dead and deadening creed. The ecstatic spirit is good, but in fleeing mechanism, Fox stumbles into the opposite extreme. Much of how human life goes on has to do with preserving things that exist, a process neither creative nor transformative in the sense he has given those terms. Preserving the existing support structure may smack of status quo-ism for certain people, or mere routine for others. Yet most people desire regular routine and predictable paths as much as plants need a regular round of day and night, and seasonal variation. Amid all the talk of revolution and ecstasy, there should be more than a little room for continuity and sobriety.

Fox at many points criticizes Saint Augustine, and with good reason, from his point of view. On the surface, this rejection stems from Augustine's putative misogyny, sexual puritanism, emphasis on fallenness, and even his lack of cosmology. Except for the last charge, all these criticisms have been directed at Augustine before by people seeking either to claim a less pessimistic view of human nature and history or to pursue a more sensuous Christianity. Fox seems to want to do both. Though he allows for the darkness in creation with his Negative Way, in fact anything even remotely negative, in the world or the spiritual life, is merely notional in Fox. The unrelenting tone is celebration, joy, ecstasy, pleasure, and fulfillment. No one, of course, would oppose any of these things properly understood. For all his own enthusiasms, Augustine, at least after his conversion to Christianity, is much more reliable and realistic about what can be expected from human nature and the world. Being realistic for Augustine is not the same as accepting the status quo. It depends on perceiving reality. The West has built on the realistic intuitions of Augustine to create everything from just-war theory to the separation of church and state. Augustine is a great spiritual and even mystical writer.[20] But Augustine was also the kind of person that you could rely upon to see that the careful discriminations and hard work of the world got done. Shall sin and its effect shall be swallowed up by the Original Blessing? Fox says yes. Augustine, much more realistically, says yes as well, but expects it only at the end of time.

There is no question—and Fox is right to emphasize it—that the earth is a word of God, spoken to us. But, as he often quotes from

Thomas Aquinas, mistakes about creatures often lead to errors about the creator. He would have done well to take his own advice. In seeking to flee an overly mechanistic and utilitarian view of nature, Fox has simply embraced an overly spiritualized and sacralized view. In trying to revitalize liturgies and spiritual practices, he has slipped vital doctrinal moorings. In creation, God provides us with many of the things we need, but nature is not a perfect garden for our bodies, or a spotless womb for our souls. Nature, including human nature, is enmeshed with cosmic struggles, both for mere survival and for blessedness. Fox is so dismayed by the Fall/Redemption pattern that he wants to wish it all away with too great an emphasis on the Original Blessing. But wishing, even pious wishing, does not make things so.

6

Sophia's World:
The Radical Imagination
of Ecofeminism

If Matthew Fox's work represents a kind of *omnium gatherum* of the more eccentric contemporary religious concerns about the environment, some elements embedded in the overall movement warrant independent examination in their own right. One of the most prominent and influential of these is a spectrum of views that is usually called *ecofeminism.*[1] Ecofeminism tries to uncover and delegitimize the forces that have allegedly gone into the construction of "patriarchal" society, in order to present a radical and sweeping critique of much religious and scientific thinking about "man" and "his" place in nature. In a way that Henry Adams once again anticipated, ecofeminism aligns itself with the Virgin, after a fashion, over against modern technological, capitalist, and democratic society, which it identifies as a largely male-centered project of establishing a rapacious Dynamo.

Of course, ecofeminism does not easily line up with traditional views of the Virgin by any means. Like the broader feminist movement from which it takes its cue, ecofeminism simultaneously carries on a critique and a reconstruction of the Western religious tradition, which ecofeminists view as having subordinated and marginalized women. For that reason, many religious feminists would try to reach behind the traditional image of the Virgin Mary as a piously submissive mother to a more vigorous, often aggressively sexual, female

185

force.[2] Quite often the redefinition of the tradition also tries to resurrect the concept of female Sophia (Greek: wisdom) from the Jewish and Christian Scriptures as an essential collaborator in the Creation, if not merely another name for the one true God. The Gnostics, too, valued Sophia, and modern feminists therefore also look kindly on certain Gnostic currents as having promised liberation for women.[3] Ecofeminism, then, involves a certain constellation of female qualities and historical beliefs in many respects quite different from traditional Western religious notions.

Ecofeminists of a religious bent frequently invoke the Gaia hypothesis as a way to recover what they believe to have been the archaic religion of humanity. In this view, the original spirituality of Europe was the worship of the Great Mother, who is also identified with the earth. Subsequent history is primarily the record of males trying to impose hierarchy, violence, and environmentally destructive forms on the naturally egalitarian, peaceful, and earth-honoring patterns earlier established by female-dominated cultures. Marija Gimbutas's book *Goddesses and Gods of Old Europe: 6500-3500 B.C.* is the primary source for these claims.[4] Like Father Thomas Berry, ecofeminists often regard the Upper Paleolithic and Neolithic cultures as superior to subsequent civilizations, all of which reflect a decline into the "logic of domination."[5] Ecofeminists' reasons for believing this differ slightly from Berry's; they believe that the archaic cultures not only were environmentally sound but also worshiped the fertility and intrinsic sacredness of the female. Human history is the record of that female paradise lost.

Besides invoking Gaia, ecofeminists also seek to identify hidden religious women's voices down through the centuries as a counterweight to the dominant historical tradition. Some may dabble — or more than dabble — in supposedly feminine alternative religions like witchcraft, which is identified with the old Mother Goddess religion. Quite often, this occurs within religious orders or female religious groups associated with mainstream Christian and Jewish communities. The modern Wicca practitioner Starhawk has been prominent in activities at Matthew Fox's Center for Creation Spirituality, and analogous practices show up at relatively mainstream events such as the annual Women-Church conference, which draws many religious women from various Christian denominations. Power crystals or pyramids, spells and incantations are not unknown in ecofeminist thought and practice. Though many of these currents may seem strangely irrational, bordering on the unbelievable in modern societies, they seem to appeal to the same broad sentiment that made New Age religions highly popular.

Five Principles

What are the principles that underlie the general ecofeminist vision? The movement is complex and could be broken down into categories that warrant entire books in themselves. But it is not entirely unjust to differences within the broad movement to say here that there are five large ecofeminist principles:

1. Female holism and organicism are basic. Because of women's intimate involvement in gestating and caring for new life, they exhibit connectedness and nurturing instincts that men, at best, weakly appreciate. The technologizing of society is the result of coldly linear and abstract male thought that tries to transcend the body and feeling. That way of thinking has led men to regard objects and life forms as distinct beings independent of one another. The proper view is to see all of nature as interconnected, in the "web of life." To lose that insight is to lose the foundation for wisdom.
2. Femaleness resembles nature. The male domination and outright "rape" of nature was legitimized by the male domination and rape of women and derives from the same sources. Our current plight is a direct consequence of the near exclusive focus on male interests. As a result, the other half of the human species must have an important say in future decisions and policies if we are not to upset the balance of nature and thus destroy ourselves.
3. Females have superior insight. Because men tend to view things abstractly, objectifying nature as mere material for use in the same way that they objectify women, they do not value the intimacy of touching and community as women do. "This paradigm, initiated through the work of Bacon and others in the seventeenth century, champions a mechanistic approach to nature in general and women in particular," writes one ecofeminist.[6] Women's direct relationship with, and concern for, living things in their environ-

ment must correct widespread male injustice towards other people and nature. Another ecofeminist writes: "Women's experience empowers them to discern the systemic and psychic links between various forms of injustice."[7]

4. Capitalism and technology are male pursuits par excellence. What is needed is not simply to redistribute power to women but to change the very notion of power within Western society. Hierarchy in human societies and a false belief that human beings, especially white males, are "higher" than other life forms are correlated with each other. From these, the sexism, racism, and militarism of the developed world naturally flow. Cooperation must replace competition; we need to repent profoundly of our sins against nature and create a bio-philic consciousness.

5. Women, the earth, indigenous cultures, Third World peoples, and the poor and marginalized in developed nations may be viewed together as an "other" that white male patriarchies, multinational corporations, and large nation-states seek to dominate or even destroy. Their combined forces must animate religious and political action if the world is to be saved.

Other principles, as we shall see, enter into this alternative vision, and particular thinkers may choose to emphasize or downplay one or another element. But on the whole, ecofeminists as a group fit comfortably within the framework briefly outlined here.

Just to state the radical feminist critique of religion and alleged Western views of nature is to invite passionate reactions. If you accept the unflattering portrait of men and believe that our environmental problems must stem directly from male traits, ecofeminism will have an immediate and overwhelming plausibility. If you have doubts about that portrait and the account of Western history to which it leads, the whole ecofeminist movement may appear entangled in self-contradictions and misconceptions that draw from phenomena having little real connection to one another and lead to large, and largely unsupported, generalizations. In particular, the assertion that men's views of women determine all their other behavior is a strong claim. Ecofeminism asks us to reread all of human life and history from its perspective, and it is therefore not amiss to inquire whether the historical claims and the proposed course of treatment are well founded.

The Immediate Context

One strong reason for doubt is the strangely disconnected attitude of ecofeminism to its immediate life situation. Ecofeminists like to boast of their connectedness with the world and their refusal to engage in abstract male analysis. Yet feminism, as they conceive of it, seems wedded to an ideology that obscures the most basic facts of their lives. Some forms of feminism are a proper response to an overmasculinized society and have led to a growing moral and political realization that women must now be treated as fully equal partners in the work of the world. But at least part of this mentality also comes to us from a curious interrelationship between the Virgin and the Dynamo: the movement for women's equality relies in no small degree on the technological aids provided by the Dynamo. Before the development of modern technologies, the physiological differences between men and women were a crucial, almost an insuperable, factor. Feminist historians continue to invoke myths of prehistoric matriarchies, but primitive societies were almost always male dominated and doubtless posed far greater obstacles to female equality than their modern counterparts.

The whole account of human societies in feminist histories tends to simplify, and at times distort, the truth even in areas that seem settled by scholarship. Most feminist historians point to societies that worshiped goddesses or that define kinship by matrilineal categories as places where women were empowered or valued. But according to the best anthropological research, matrilineal systems, which have existed in many cultures since the Stone Age, are only a system for classifying how people are related; they should not be confused with matriarchies, societies in which women are the rulers. The eminent anthropologist Claude Masset dismisses such notions:

> I will say nothing of manifest improbabilities, such as "primitive matriarchy," which was part of Marxist thinking ever since its illustrious founder borrowed the notion from the early ethnologists. Not only do our closest cousins, the gorillas and chimpanzees, have social organizations the core of which is always a male or a group of males, but of the ninety species of primates there is not a single one in which females exercise authority within the group. In order to find populations normally led by females, we

have to turn to European deer or bison, which, to say the least, are not among our closest relatives.[8]

His colleague, Françoise Zonabend, is even more emphatic about the human record. She observes: "Matriarchies, those societies said to be governed by women that some authors in times past equated with matrilineal social systems, claiming that descent was passed on through the female line, have never existed, except in the mythical memory of societies or in the imagination of the first social anthropologists and historians of family law."[9]

It is true that the human race, even in its primitive or tribal forms, displays a wide variety of social and gender arrangements, and each of these may have advantages and disadvantages. But men's superior upper-body strength and its central importance to societies whose survival depended on their being able to defend themselves (virtually all societies except for small isolated groups over brief periods) made it virtually inevitable that women would take up less prominent roles. Further, the sheer backbreaking domestic work that needed doing before modern appliances were invented (and that men in primitive societies, like their contemporary counterparts, generally shunned) led inevitably to women's being assigned onerous chores, while men were often called on to undertake defense and what we would call today political activities. Some writers have tried to portray agricultural societies, where much work was shared and performed in or around the home, as communities of greater equality between the sexes. That is partly true. But the historical record still seems to prohibit any easy belief that premodern societies were better for women.

In the most concrete sense, equality for women has a technological as well as a moral dimension. Machines have made all forms of work far easier. And advances in medical science have turned pregnancy and delivery into far less dangerous events than they were in the past. For those and other reasons, women's life expectancies, which were far lower than men's in premodern societies and remain so in developing countries, are significantly higher than men's in the developed world. Modern economies, too, which rely far less on human muscle power than any previous social organization in history, have created a myriad of jobs that women, however much they may differ from men in other respects, are perfectly capable of handling.

The freedom of women in modern societies is greater than in traditional societies and may call for some counterbalancing so that women who want to stay home with children are not stigmatized for not pursuing careers. Modern society, despite the imperfections that always exist in every human community for male and female alike, is beyond all question a great blessing for women.

The Self-Contradictions of Ecofeminism

Because it is not aware of its context, ecofeminism exhibits many internal contradictions. For example, "taking control of your life and reproductive power," the phrase often employed by feminists of all kinds, ecofeminists among them, contains a curious modern contradiction. Control is usually a negative term in feminist thought, associated with a patriarchal mode of domination that the feminist theorists are quick to reject. In ecofeminist writing, even self-control is frequently characterized as a masculinist assertion of reason over feeling. Ecofeminists regard control, especially when conjoined with technology, as a toxic notion. Yet those same women have no qualms about using various contraceptive technologies to control one of the sacred dimensions in ecofeminist thought: female fertility. Indeed, they are not much concerned about the violent control and domination that abortion exerts over vulnerable life. If nurturing and connectedness were as hard-wired into the female body as ecofeminists suggest, it is doubtful that they would promote either of these practices.

This issue reflects a larger problem with ecofeminism. Just as ecofeminists seem disconnected from the contemporary conditions that make feminism possible, they are curiously disconnected from specific environmental questions per se. Ecofeminism is rich in abstract theories about male domination of societies and nature. It is almost wholly lacking in any engagement with ecological science, biology, or the other disciplines that are needed to understand and redress specific problems. One explanation for this lacuna may be that ecofeminism has so thoroughly swallowed certain forms of feminist theory that regard science and rational analysis as male modes of domination that it badly undervalues or simply dismisses them. But without the scientific breaking down of nature into discrete, understandable pieces, which are later reassembled to permit an understanding of an interactive biosphere, real environmental work cannot be done at all. Mere criticism of oppressive patriarchy borders on a self-absorption that relegates the real problems to a secondary realm, one that apparently will be easily managed once sexist ideology is corrected.

Women, of course, are quite capable of doing serious ecological work and are doing it all the time. As we saw in chapter 2, Maureen Raymo of MIT has been the preeminent researcher into the geological history of a pressing current concern: global climate change. But unlike the ecofeminists, who lazily posit that nature must be a stable harmony and that intuitive relations trump all other considerations, Raymo has discovered that global climate is far from stable and harmonious, having exhibited many wide and rapid swings caused by unknown forces in the past. Nor did she arrive at this conclusion on the basis of some previously concocted ideological notion of intuitive connectedness with Mother Earth. She did the science and got results that are credible to everyone, male and female, who believes evidence should take precedence over comforting illusions. Furthermore, women who do ecological science generally refrain from the facile assumption that once you identify the problem — say the earth's capacity for quick climate change — you know the solution. Solutions, too, take hard work even after you have a fairly good idea of potential problems. Many women are engaged in valuable efforts to come to a real knowledge of the world in which we live; ecofeminism seems to regard them as part of some male science that by its nature threatens the earth. But in a more balanced understanding, there is no credible alternative to scientific analysis and action. All the gender egalitarianism in the world will be of no use to us if, say, we face man-made or natural global climate change.

Feminist Critiques of Ecofeminism

Fortunately, some countercurrents from within feminism itself may open up new and more effective women's perspectives on the environment. For instance, Janet Biehl, who approaches environmental questions from a radical Left perspective that directs strong criticisms towards capitalism, nation-states, militarism, and ethnocentrism, rejects the facile ecofeminist invocation of the return to goddess worship as a solution to ecological problems. Biehl concedes that the old European communities, often portrayed as ideal in ecofeminist literature, may have been more egalitarian, less warlike, and better environmentally than the average, but she disputes the idea that the admirable elements in these poorly understood groups are the result of goddess worship. For Biehl, this is a monocausal simplification similar to Lynn White, Jr.'s "very naive judgment" that Western ecological misdeeds should be blamed on the Bible.[10]

Instead, Biehl thinks that the relatively simple and harmonious Neolithic social forms may be the cause of goddess religion rather than the reverse. She cites several prominent scholars who deny that there is any particularly strong correlation between goddess worship and the high status of women. The classical specialist Sarah Pomeroy has sharply observed: "The mother may be worshiped in societies where male dominance and even misogyny are rampant." Low status for women, it appears, can exist even where religious hierarchies include priestesses. Biehl cites the worship of goddesses in China that coexisted with the well-known Chinese denigration of women and the high status of the Virgin of Guadalupe in Mexico despite the macho character of that nation. These and other documented cases from historical times lend little support to an ecofeminist reconstruction of prehistoric societies that, by definition, provide little evidence. Such evidence as there is requires much speculative interpretation. It is perhaps telling that Riane Eisler, who has popularized ecofeminist history in *The Chalice and the Blade,* describes those ancient societies — so distant from us in time and attitudes — as "structured very much like the more peaceful and just society we are now trying to reconstruct."

But Biehl also doubts that these Neolithic ideal communities were quite so peaceful and environmentally aware as is believed. Both attributes, if they really existed, may have had more to do with the relatively small size and isolation of the groups. We just do not know enough about them to take them as a historical precedent. The whole ecofeminist historiography is, for Biehl, an ideologized oversimplification. For instance, the usual explanation for the eventual disappearance of the small Neolithic communities proposed for our imitation is that "Kurgan invaders" from the Russian steppes imposed male dominance and violence on otherwise mild people. But whether there even were any "Kurgans" or any invasion is still uncertain. Biehl thinks, on the basis of known cases of social development in historical times, that the primitive hunter-gatherers may simply have evolved from within to more highly articulated societies that resembled the so-called Kurgans. Furthermore, the sites that have been studied at Obre (Croatia) and Vinca (Serbia) were described, even by Marija Gimbutas, as possessing worrisome traits: "human sacrifice accompanied by animal sacrifice was performed in open-air sanctuaries,"[11] and these sites include evidence of the ritual sacrifice of small children. As with the Babylonian mythology of the female serpent Tiamat and the Meso-American "blood of the gods," what we know of religions in historical times gives no reason to think that such things are impossible.

But all this is of little interest to most ecofeminist proponents of the Old European Goddess. In fact, much of subsequent history is equally ignored. Everything after the arrival of the Kurgans is merely a record of male warrior domination and hierarchy that look on nature as dead matter for male exploitation. Biehl herself criticizes a good deal in subsequent developments, but she does not think much of the ecofeminist equating of all hierarchy with patriarchy: "priestesses . . . somehow seem to be an acceptable form of hierarchy to ecofeminist-theists."[12] And she says of the division of all human history into Before Kurgan and After Kurgan:

> So much for all literary, philosophical, technical, musical, artistic, and cultural achievements of the millennia between the disappearance of the idyllic Neolithic and the modern era. Ironically, even a passing reading of Western philosophy will show that "post Kurgan" times were resplendent with richly endowed organismic philosophies and cosmologies, indeed, that a "mechanized world view" was a very recent phenomenon.[13]

The New Victorians

In a slightly more popular vein supported by good scholarship, Rene Denfeld has characterized these same ecofeminists and other censorious and self-righteous feminist figures as "the new Victorians." In her book by that name, Denfeld disputes some of the same points in the ecofeminist orthodoxy that Biehl does, but she adds that the moral arrogance, megalomaniacal claims, and outright unreason of much ecofeminism are turning young people away from what is good in feminism. Describing a West Coast convention for Women's Day, she gives a virtual catalog of how the goddess myth, shaky in its historical foundations, takes on quite aggressive forms at the popular level. She notes that at the conference "table after table [was] arranged with colorful assortments of goddess statuettes, crystals, feathered Native American pipes, rich wood sculptures, necklaces bearing little pagan charms, and other New Age spiritual apparatus. Flimsy paperback books on the 'goddess' are offered for sale alongside mystifying hand-painted sticks and small stone bowls. There are calendars that feature different goddesses for each month and ritual books on modern witchcraft."[14] These items, of course, have become the staple of feminist and many religious women's events in recent years.

But perhaps even more striking than the vaguely irrational, sometimes explicitly magical objets d'art of the dominant mode of ecofeminist social gatherings, says Denfeld,were the "facts" the local National Organization of Women (NOW) chapter attributed to the goddess religion:

> There were no weapons of violence, no fortifications surrounding goddess communities and no warfare.
>
> The earth was, and is, the goddess' body. The earth is a living, breathing entity and all life forms are sacred.
>
> Women controlled the religious, social, political, legal institutions, which were all focused on community improvements and equitable distribution of wealth and work.
>
> Every woman was required to work as a priestess for varying amounts of time each year, men were on all important community counsels, though women were given the highest honors.
>
> Women were free to change sexual mates, bear or not bear children, without social guilt or belittling stigmas attached to their actions.
>
> People were closely connected to nature, its cycles, its power and beauty.[15]

Denfeld comments tartly: "It sounds like a feminist Disneyland."
Probably all complex historical notions undergo some oversimplification when popularly presented. But the obvious tailoring of a few
archeological hints to a whole modern ecofeminist agenda may reach
something of a high-water mark for creative adaptation. Nor do
women have to go to feminist conventions to find these oddities.
Many colleges and universities now teach them as established truths,
often with little concern about whether they are advocating specific
religious and political agendas in the classroom. At the University of
New Hampshire, for example, these notions are taught in a course
that ends with a "closing ritual."

Affirming Intuition

Outright magic plays no small role in a large part of the
ecofeminist movement, and Denfeld worries about the ways in
which the superstitious, antichurch rhetoric of ecofeminism may
drive women away from work that needs doing. The Wiccan
Starhawk has claimed that a magic circle she arranged on a hillside
above the Diablo Canyon (California) nuclear plant led to its closing
(it later reopened); she also believes her spells led to the elimination
of Pershing and cruise missiles in Europe. Quite often, these practices mix up Neolithic notions, Native American cultures, and East
Asian and aboriginal beliefs, ignoring the differences—or sometimes
contradictions — among them. Female biology is tied to pagan respect for seasonal cycles, and feminine intuition to women's superior
ways of knowing, and these are supposed to offset the usual rational
crudity of maleness. That biology as destiny and women as incapable
of rational thought were nineteenth-century notions used to oppress
women, says Denfeld, are unknown to these historical innocents:
"Like Spencer, goddess feminists do not base their ideas of female intuition—or for that matter, any other supposed gender differences—
on any even faintly scientific evidence. They don't have to. They just
have a feeling."[16] Identification of women's wisdom with menstrual
cycles and fertility, she also fears, may eclipse their other capabilities
and contributions.

The innocence of the goddess societies also does not stand scrutiny. David Anthony, an anthropologist, has argued that far from

showing a society without conflict, some of the Neolithic goddess sites show "massive fortifications." Denfeld remarks that it was a slander on the Neolithic people to view their lives, as Hobbes described them, as "solitary, poor, nasty, brutish, and short," but the phrase is not entirely wrong: "Few lived past the age of forty. People faced starvation over long winters and in times of drought, women in particular faced the dangers of childbirth without modern medicines, and children often died from disease. And apparently many faced warfare and death by skull-crushing maces, along with the chance of ending life as a sacrifice."[17]

The myth of early goddess societies seems to be exploded for a variety of reasons. But what of the ideal vision that projects back on the past values that we hope to realize in the present? Cannot those values survive a mistaken attempt to find them somewhere in the past? The answer to that question is both yes and no. Yes, when the ideal corresponds to some reality; no, when the ideal itself has been corrupted by a vision of human life that runs contrary to nature. Though there is a relatively large group of more sophisticated ecofeminists schooled in theology, philosophy, and history — some of whom dispute the facile mythmaking of vulgar ecofeminism — it would be difficult to say that those ecofeminists escape self-contradictions as fatal as those found in the thought of their less intellectual sisters.

Holism and Organicism

For the sake of clarity, it may be useful to see how the sophisticated ecofeminists handle the five principles we laid out at the beginning of this chapter, beginning with notions of holism and organicism. As Janet Biehl mentioned, it is not difficult to find many examples of both of these concepts, along with many other remarkable views, in the alleged dark ages of patriarchy. What, then, might distinguish ecofeminist holism and organicism from the Western male variety? Certainly one of the claims often made by the religious ecofeminists is that the dualism God/nature conceived of by various male philosophers gives rise to a cascading series of other dualisms. Grace Jantzen, making a near universal, ecofeminist argument, not only sees the cosmic dualism of God and a world created ex nihilo as fundamental to all other dualisms (mind/body, male/female, master/slave, thought/feeling)—that in itself would take no little work to prove—but goes still further, in a fashion that is typical of most ecofeminism, charging that the cosmic dualism was deliberately constructed

> as a theological justification for patriarchy. The dominant group of ruling class males constructed a world-view which set them apart as normative humanity, over against the "other"—women, other races, the poor, the earth — and then fashioned in their own image a God of ultimate value, power, and rationality over against the disvalue, passivity and irrationality of the opposite side of the duality.[18]

There is no clearer—and more unbelievable on its face—formulation of the way in which some ecofeminists would try to see all human history through a distorting lens.

More serious for the deeper question about holism and the environment are the specifically religious principles that some writers try to draw out of ecofeminist metaphors and analogies. Fairly typical in this regard is theologian Carol P. Christ, who pits feminist images of the earth as our organic place of origin and home against some forms of rationalist Protestant theology of a decidedly Kantian bent. These theologies emerged primarily during the period when Newtonian physics was in the ascendancy and the unique moral and spiritual status of human beings seemed to consist in their radical separation from the rest of nature, particularly the naked moral will.[19] Kant and his followers sought the essentially human in pure reason and moral decision apart from the influences of emotion; feminists like Susan Griffin, by contrast, remind us that "we have cause to feel deeply."[20]

For Carol Christ, this opposition reflects a deeper conflict between "the language and thought forms of patriarchy . . . the voice of male philosophy and theology" on the one hand and the voice of "our connections to the powers of beings" on the other. But the question is quite a bit more complex than this formulation suggests. As we saw in chapter 1, a traditional theological figure like Cardinal Ratzinger, too, wishes to emphasize the rootedness of our creation in the earth. The kind of critical philosophy and theology that Carol Christ identifies with the whole of the tradition has long been thought problematic within that tradition itself, sometimes for the very reasons she finds in feminist insights. But unlike Carol Christ, those theologies thought it important to preserve specifically biblical insights into a larger rationality inscribed in nature. The eclipse of, and downright opposition to, a truer rationality not reduced to abstract logic and including science represent a blindness in ecofeminism that may overwhelm its insights.

Carol Christ can easily show that the kind of poetic evocations of nature in various feminist theorists is very rich fare compared with the thin speculations of rationalist theology. But that theology itself has always believed that God and reality are beyond ultimate naming; only extreme scientific forms of theology have ever thought that theologians could do for God what natural scientists hope to do for nature — provide a total explanation. Poetic, Native American, and ecofeminist "communing" with nature is valuable testimony to the human spirit as it encounters reality, but Carol Christ wishes to erect on these feelings a universal valuing of all things that would be difficult to square with the biblical vision or our everyday experience.

The most minute aspect of creation is infinitely beyond our power to know and value, but the biblical vision still reserves a special place for human life amid that divine embarrassment of riches precisely because we seem to be the only species that responds with awe and admiration to the universe. Or as the other Christ, the carpenter from Nazareth, put it: "See how the lilies of the field grow. They do not labor or spin, yet I tell you not even Solomon in all his splendor was dressed as one of these" (Matthew 6:28-29). Somehow Jesus did not see any contradiction between this truth and something he had said just a moment earlier: "Look at the birds of the air; they do not sow or reap or store away in barns, and yet your heavenly father feeds them. Are you not much more valuable than they?" (6:26). A little later, he even puts it mathematically: "Don't be afraid; you are worth many sparrows" (10:31). And all this is connected by Jesus with great responsibilities.

Carol Christ is dismissive of these dimensions of the Bible because they speak to our existential sense of human life as encompassing things beyond nature, while she thinks any such claim is dangerous in tempting us toward abstract transcendence. For her, valuing "the color purple," in novelist Alice Walker's phrase, in which God is "pissed off" if we pass a field of that color without appreciating it properly, takes precedence and excludes the rest. She would prefer a purely celebratory, joyous, and diversity-appreciating humanity to the old Christian balancing of prerogatives and duties. In part, that may be because she appears simply deaf to the asceticism that is a part of every serious religious life, which must grapple on the path to liberation with issues of self-discipline within freedom. In particular, she believes women have simply been deluded by such talk of self-discipline into "service to God."

Christ regards the Jewish conceptualization of God's rule over man and nature as the basic covenant to be found in the Bible and parallel to other forms of hierarchical domination. She finds emphasis on the Crucifixion as a kind of paradigm of "self-giving" dangerous, especially for women. She even wonders whether "this image of Christ crucified can be separated from the other, more dangerous image of Christ exalted to the right hand of the Father," which she associates with male imperialism and a certain platonizing belief in transcendence of nature.

For most Christians and Jews, of course, these are the issues that go to the heart of their experience of the moral and spiritual struggles of life. Indeed, the kind of creature who can know and wrestle with such questions always seemed something extraordinary in nature. Carol Christ rejects all such notions, because in her desire to preserve our consciousness as a part of the "web of all life," in which each thing has its own infinite, nonconvertible value with respect to other things, she loses other valuable insights. There is, of course, a grain of truth in her view, but it is the kind of half-truth that theology has always had to balance off against others.

It is in theological work like this that we come directly up against whether we will remain within the biblical orbit at all if we take ecofeminism as normative. For thinkers like Carol Christ, it would be no great loss if we simply forgot about the whole Fall–People of God–Redemption story of the Bible. Her real concern is a kind of cosmic mysticism in which human beings find themselves neither more nor less valuable than any other creature. The mysticism is an intriguing vision, and one that has often in history found a niche within the biblical view, properly understood. But in terms of helping us to understand our *environmental* problems, and more especially in their capacity to tell us what to do about these problems, religious refashionings of the tradition like this one have little to say and may actually do much harm if they encourage inattention to real nature, as well as to the human thought and creativity that also seem an established feature of both the natural history of humanity and the biblical history.

Femaleness and Nature

The second basic principle we saw was the similarity of femaleness and nature, an old claim in many cultures. For years, feminists fought this identification as demeaning to female thinking and freedom. With the appearance of ecofeminism, however, it has become a badge of honor. The analogy between male domination over women and domination over nature, however, begins to appear flimsy the moment we look into it more carefully. By definition, an analogy exists between two things that are both similar and different, with the difference the more prominent part of the relationship. All dominations have similar elements, but does the recent technological ascendancy over nature really mirror a devaluation of women? If that were so, we would expect women's position to have deteriorated since the scientific and industrial revolutions. But in spite of the many ways in which modern societies are uncongenial to women, that is precisely the opposite of the indisputable historical record. Those revolutions have been liberating for women as for men. Sometimes they seem even better for women.

It sometimes appears that ecofeminists are arguing that male scientists or capitalists are at heart rapists in the literal sense. If there were a strong correlation between rapacious male intellect and will and mistreatment of females, by now there would be a whole literature confirming the greater incidence of female battering and rape among those who practice these male perversions. But no such literature exists, and the whole notion that mistreatment of nature originates in mistreatment of women appears nothing more than an ideological fiction. Many businesspeople and scientists, concerned as they are with both preserving natural value and respecting women — when they are not women themselves — might take strong issue with such a view.

One of the more powerful, if radical, ecofeminist theologians, Rosemary Radford Ruether, has disputed the oversimplified history that "reduces everything to one drama: the story of original female power and goodness, and the evil male conquest and suppression of the same." In her book *Gaia and God,* Ruether points out that the goddesses of Middle Eastern cultures supported male hierarchies.[21] Similarly, she disputes the notion that witchcraft, or Wicca, was historically a religion of a goddess.[22] But her scholarly work and activism, for all the knowledge of theology, philosophy, and history that she brings to the subject, is not far distant from the positions of mainstream ecofeminists, and she actively encourages pursuing the ideal vision even if its historical basis is false.

For instance, despite her scholarly reservations about the simplified historiography, at one feminist event she argued baldly that "the psychosocial root of our ecological problems is sexism, the subjugation of women." This often repeated but never much explained notion led to another: "We are parasites, utterly dependent on the rest of the food chain. Nature would be much better off without us."[23] A thinly concealed misanthropy and curious harshness emerge here from a writer who generally identifies exclusion or demotion of any group as a quintessentially male vice. Almost every living thing depends in some fashion on other living things for food, processing of chemicals, and a variety of other functions. Does that mean that all living things are "parasites" on the earth? Would Earth be better off without them too? And what would "better" mean in such a context?

Elizabeth Johnson provided a slightly better formulation in a paper she gave as she finished her term as president of the Catholic Theological Society of America:

> In an interdependent system, no part is intrinsically higher or lower. Yes, more complex life represents a critical evolutionary breakthrough, but not such as to remove humanity from essential dependence upon previously evolved creatures. The challenge is to redesign the hierarchy of being into a circle of the community of life. With a kind of special humility we need to reimagine systematically the uniqueness of being human in the context of our profound kinship with the rest of nature.[24]

This is better not only because it does not see us as uniquely parasitic but also because it tries to preserve both human uniqueness and a sacred sense of the world. But in the same paper Johnson falls into the usual facile and unexamined ecofeminist rhetoric: "To be truly effective, therefore, the turn to the cosmos in theology needs to cut through the knot of misogynist prejudice in our systematic concepts, shifting from dualistic, hierarchical and atomistic categories to holistic, communal and relational ones." But not every dualism, surely, is vicious (Johnson's own notion of interdependence divides even as it unites). Hierarchies and atomistic notions are useful for some purposes. And coupling all dualism with misogyny requires a great leap of faith.

There is a great deal to be said for recognizing our interdependence with the rest of nature, both as a spiritual practice and as a practical principle. But this interdependence cannot be applied unimaginatively. The human species represents the complex, developing side of nature. The human brain, for example, is the most complex structure of which we are aware in the entire universe. Now that technology has enabled us to meet many of our basic needs, we will also wish to factor into our thinking many truths about the complex web of life in which we live. But that points not so much to some sort of willful denial of our own powers as to a further development in human consciousness that includes our awareness of the threats as well as the benefits of our technology.

Superior Female Insight

Though ecofeminism is, in principle, a movement towards equality between the sexes, in practice it privileges what are thought to be superior female insights. Marti Kheel, for example, has written lucidly about the differences between ecofeminism and Deep Ecology, a mostly male movement.[25] For Kheel, Deep Ecology's call for a "new consciousness for all of life," in which notions of the Self are expanded so as eventually to include all beings within the person's own wish for self-care, still remains within a masculine perspective. For her and many ecofeminists, the basic problem of Western culture and the environmental issues to which it has given rise cannot be solved through this wider identification. The West has been marked by an anthropocentrism that tries to identify life with us rather than identifying ourselves with life, a more properly female view. She locates the specific source of the blindness in *androcentrism,* the privileging of male perspectives.

Kheel doubts that women experience themselves or the world as the apex of nature the way men do. The acceptance of hunting, for example, by some Deep Ecologists as a natural and age-old activity that must be undertaken in full consciousness of the deep mutual relationship of hunter and prey worries her. It indicates that Deep Ecology retains a perspective of the male need to prove oneself against the natural world that women never, or only rarely, feel.[26] Women, as Charlene Spretnak has argued, have other means of experiencing a oneness with nature in "menstruation, orgasm, pregnancy, natural childbirth and motherhood."[27] Kheel provides a sensitive and lucid account of why men, constrained as they are to establish an identity separated from the mother while retaining a need to return to a natural unity, love hunting. But this overlooks the ways that women too must find identity separate from their mothers.

Her analysis might have benefited from a wider view of men and women as part of the very natural creation for which she has such concern. If men, by their very nature, are inclined to different activities than women, an ecofeminism that bases itself on the hope that men will simply become women has cut itself off from one half of the human race and an important part of nature's own development. There may be good reasons in nature why male inclinations to face dangers, invent new ways of living, and seek a kind of independence are good for the human race and nature. In Kheel and other ecofeminists, an abstract oneness with nature, regarded as a kind of original or normative position, excludes much of what must constitute human flourishing. Human life in this perspective ought to be lived more like a vegetarian picnic than the challenging and creative process we experience it to be. In that regard, the Bible's firm assertion that "male and female he created them" might serve as a corrective for a secular feminism that at bottom seems committed to seeing only one half of the human pair flourish. Such attempts must meet the charge of trying to feminize the human male.

Male Capitalism and Technology

Nearly all ecofeminists regard the marriage between technology and capitalism as evil. Capitalism for them means cutthroat competition and the dominance of greedy industries over our connection with nature. Unlike Father Thomas Berry and his "planetary socialism," ecofeminism is weak on ideas of what social and economic forms will replace rapacious capitalism. And few even have a very good idea of what capitalism contributes to ecological problems. Elizabeth Johnson, in the essay cited above, for example, seems to believe that development in Third World countries leads only to "deforestation, soil erosion and polluted waters";[28] of the benefits of development, she speaks not a word. We have noted above that ecofeminism generally pays little attention to concrete environmental issues. When it does, it often falls into predictable clichés.

One exception to this rule is the strong global critique of capitalism that Rosemary Radford Ruether presents in *Gaia and God*. Ruether boldly dedicates a book intended to be a scholarly examination of the deep theological and philosophical roots of Western environmental degradation to people killed by American bombs in Baghdad during the Gulf War. The implication is clear. The West's addiction to cheap fossil fuels, particularly Middle Eastern oil, is one of the roots of its militarism. And the militarism itself is in defense of the exploitative and wasteful consumerist culture, the American way of life, that we are committed to preserving at, almost literally, all costs. Ruether sees the demilitarization of America and Israel as the necessary first step in establishing a new global ecological order.[29]

Like many ecofeminists, Ruether prefers cooperation to competition. Unlike them, she rightly distinguishes competition within human societies from competition for resources in the animal and plant kingdoms, where no ethical principles and a kind of rough balance exist. But she inclines to the view that economic competition is a wild, unguided vehicle for the "annihilation" of the opponent. Like ecofeminist ecological science, ecofeminist notions of economics are quite rudimentary. Elsewhere, Ruether has suggested that economic growth is a kind of zero-sum game, in which women and minorities in the First and Third Worlds have been "impoverished" by the way others have grown wealthy. But this does not withstand serious scrutiny. All societies were once "impoverished"; there is no need to see poverty as a product of success elsewhere, unless all economic development is banditry. What needs to be explained is how wealthier societies got that way. Exploitation of natural resources and weaker groups is not much of an explanation.

Because Ruether at least tries to outline an alternative to industrial capitalism, it may seem unfair to criticize her for something many others in the ecofeminist movement have not even attempted. But as in most fields, a few figures set the tone and others largely follow. In addition to the usual complaints against capitalism and the environmentally damaging technologies that she believes inevitably stem from a rapacious mindset, Ruether recommends the usual soft alternatives. We should encourage small farms, try to eat lower on the food chain, save rain forests, restrict long-distance food transportation to promote bioregionalism, and generally come to know biorhythms better. Population has to be held in check. Solar, wind, water, and other alternative energy sources should be promoted, especially since our fossil fuels are limited. As desirable as some of these aims may be, it is unlikely that they will suffice for the human populations that exist. In fact, as we have seen, some of them would place greater burdens on the environment. Intensive farming saves more land, mechanized farm machinery requires fewer acres to be cleared for pasturing domesticated animals; the list goes on and on. The limits that Ruether finds in natural forms are merely the limits that once existed; today entirely new technologies will have lower impacts while respecting other limits than the technologies of the past.

It is merely a built-in bias, for example, that Ruether does not recognize the potential of nuclear energy, including fusion, to solve many problems. Nuclear energy, of course, also presents some risks and does not suggest the kind of nostalgic "natural" sourcing of wind and water and sun. But that technology, especially as a generator of clean electricity in the coming decades when human demand will increase sharply despite ecofeminist calls for cutting back use, might be one of the most earth friendly of all, especially since it is suggested to us by the nuclear reaction that has dominated the energy of the earth for billions of years, the sun. Ruether has a nostalgic bias for rural over urban life. But urban life, properly reformed, promises to have less overall impact on land as the areas that need to be built up for human use are concentrated to produce economies of scale.

Like environmentalism generally, ecofeminism's basic vision of the future is a reduced human prospect. It lacks, one might say, a certain masculine hardiness and entrepreneurship. Like many of the figures we have examined earlier, the ecofeminists seem to believe that simpler, older lifestyles are more suited to earth and will have a smaller impact on nature. It is more plausible that our own creativity and dynamism were given to us precisely so that we could limit our impact on earth by developing cleaner, stronger, smaller, integrated technologies plus the economic incentives to enable us to get them into as many human hands as possible. Capitalism has done evil things; technologies have been dirty. But as the old theologians use to say, *abusus non tollet usum,* abuse does not take away the use. Ecofeminism is too occupied with other matters to have much to say about such details.

The Question of the "Other"

Ecofeminists try to tie in a myriad of groups that are clearly political allies in a metaphor of an alternative "Other" that must oppose patriarchal ecocide. Drawing on Hegel and Sartre, Simone de Beauvoir, in her often cited *The Second Sex,* was the primary theorist of the notion of woman as "other."[30] Hegel's Master and Slave dichotomy, and Sartre's idea of the "look" of the other as constitutive of self, made the struggle with others part of what constitutes each of us as an authentic person. Beginning from this view, ecofeminists catagorize people of color, homosexuals, Third World cultures, and others as one large phenomenon opposed, like the Jungian "shadow" or a bad conscience, to Western hegemony over politics and the environment. In some instances, the very origins of modern science and technology among First World, white, middle-class males evoke suspicions that ecofeminism needs to expose the "gendered" nature of science.[31]

But there are many false analogies and mixed metaphors in this conceptual coalition as well. Most Third World or non-Western cultures show nothing like the respect for women or nature that this radical feminist vision would like to claim. In all premodern societies, female deities or principles appear as part of local mythology or an elaborate theological system. But that does not mean that real, existing women in those societies had very much influence on the overall social practices. Nor do we find ideal ecological behavior in African tribes, among Native Americans, or among aboriginal peoples. Ecofeminists are simply painting with a broad brush here for the sake of political effect.[32]

Nonetheless, the Third World provides a useful foil for criticism of the First World. Again, Rosemary Radford Ruether puts the case most forcefully in her introduction to *Women Healing Earth*. She agrees that some psychospiritual practices

> can become a recreational self-indulgence for a privileged Northern elite if these are the *only* ideas and practices of ecofeminism; if the healing of our bodies and our imaginations as Euro-Americans is not connected concretely with the following realities of over-consumerism and waste: the top 20 per cent of the world's human population enjoys 82 per cent of the wealth while the other 20 per cent scrapes along with 18 per cent; and the poorest 20 per cent of the world's people, over a *billion* people — disproportionately women and children — starve and die early from poisoned waters, soil, and air.[33]

The unstated implication here is that the poor are poor *because* the rich are rich, a notion long discredited in secular development circles. Further, Ruether suggests that the exploitation of developing regions also results in pollution from (presumably multinational) corporations. This, too, gets all but a tiny part of the dynamics of development exactly wrong.

Many of the deaths Ruether—and everyone—deplores have "natural" causes. Development would help with these problems. The Third World women cited to support opposing views often seem to have been chosen because they echo positions already developed by First World feminists. In other contexts, Ruether has recognized that non-Western views are not to be regarded as good simple because they are outside the dominant global culture. But she looks for light in Hinduism, Taoism, Buddhism, and Confucianism and places great value on pre-biblical Near Eastern cultures and non-Western traditions like the Mayan, Incan, and Aztec societies because "the three monotheistic faiths of Judaism, Christianity, and Islam were incorrect in rejecting polytheistic 'nature' religions as simply false and evil."[34]

In a Different Voice

In *The Feminist Question,* the best theological analysis of religious feminism that has appeared in this country, Francis Martin, a biblical scholar, puts all such issues into a different perspective.[35] At bottom, says Martin, religious feminists tend to make a fatal mistake characteristic of modern thought. With the rise of modern science and technology, causation has been viewed as a type of domination. A God who forms the world, it thus appears, can only be a kind of dictator who controls all things. The premodern biblical tradition saw things quite differently, and in some ways in harmony with the ecofeminist view. In the Bible, Creation is not primarily an act of domination or even a configuring of the scientific laws that define physical processes but is rather a generous outpouring and gift of being. Included in that gift, at least for human beings, is a profound freedom and even independence. Ultimately, of course, that freedom and independence will find their full flowering in recognizing God's order. But nowhere in this vision is God conceived of as dictator, much less as a patriarch, as feminists use that term.

Also included in this vision is a far different analysis of the question of transcendence versus immanence. Because of a false modern sense of these terms, argues Martin, feminists (and, it might be said, theologians like Matthew Fox) view the old notion of God's transcendence as meaning he is merely absent from the universe. A God that distant, perhaps one who set the whole cosmos going but can have no further effect on it, is an unreal God. But, says Martin, the older notion was that God could be present everywhere precisely because he was transcendent. To put this in terms of new scientific notions, everywhere in the universe is the site of the big bang because the originating singularity expands to create each point in space and every event. Thus, it is only because God transcends, and is not exhausted or defined by, the laws of the cosmos that he can be Creator, Lord of History, Redeemer, and even miraculous intervener at specific moments. The God who cannot be named in Jewish Scriptures because he transcends all categories may also be locally present as the second person of the Trinity.

Martin also has no problem with natural hierarchies. In fact, he does not see a contradiction between notions of equality and notions of ordered rankings. The king in ancient Israel has many privileges because his primary role is to be a caregiver to the people. The whole history of the Jewish people is full of stories in which the youngest son, the least likely figure, David, is raised to kingship, while kings who misuse their position are brought down. Applied to modern democracies, these ideas might be translated as preserving not only the equality of all before God and the law but also the notion that good social order demands that some be chosen as leaders and invested with special powers over others. When the leaders use these properly — that is, both morally and within an accepted juridical framework — the necessary order and authority that result enable, rather than limit, human freedom.

In Martin's thought, a radically feminine view of the depth of the problems we face does not depend upon a radical rejection of the Western religious and intellectual tradition. The great modern Swiss theologian Hans Urs von Balthasar saw the fundamental religious task in the twentieth century as a need to recover what he termed "Marian" spirituality. In his view, a Marian-inspired theology does not merely return to some pious forms familiar to us from the past. It must also recover the notion that, for the Christian person, the very first movement of piety must be modeled on the Virgin Mary's passive hearing of the Word spoken to her and her subsequent acceptance of that address. Everything else in the Christian life stems from a feminine receptivity on our part where the initiative and controlling paradigms are God's. That receptivity must be a full receptivity to reality as it may be apprehended, not only in mystical contemplation, but also in the knowledge we achieve by attending minutely to the world in its immediacy and in our scientific and technological extensions of the initial reception. Passivity and activity, connectedness and creativity are part of the full religious response to the world.

7

Liberation and Its Discontents

A mong the religious environmental thinkers we have examined in previous chapters, by far the large majority have in the back of their minds an image of human perfection and liberation. To put this image into biblical terms, they usually seek a return to the Garden of Eden. The whole thrust of modern society away from immediate contact with nature and towards an uncertain and troubling future has left all of us scrambling for sure points of reference and a direction that will reduce anxiety about the harm that the industrialized and technologized Dynamo may do and in several cases has done to our physical and spiritual environment. Since it is difficult to form an image of a better future along the general path the race has taken, many therefore seek liberation from our current situation in an idealized past.

The argument of this whole book is that, in religious as well as scientific terms, a return to Eden is impossible for several reasons. Environmentally speaking, at least, there never was an Eden. No one living today would find the physical world of the earliest humans anything other than daunting. To confuse a state of spiritual perfection, in which the race was in direct contact with God, with a state of material perfection does not help us to think clearly today. Whatever some modern theologians may think about the emphasis on the Fall in Western thought, the command to be fruitful and multiply and have dominion precedes the Fall in the biblical story. No civilization we can imagine as human can do anything other than choose wisely how to use the resources built into the world by God. Otherwise, we

211

must assume that the world would have provided everything we need without human intervention, something revelation and scientific knowledge about the direction of nature seem to show is simply false. The image of man in nature as plucking a few fruits, living quietly, and generally not disturbing, or being disturbed by, other beings is an illusory paradise. We were built by God and nature for no small number of struggles, and much more than that.

What, then, might the Bible tell us of the human adventure? If we are looking for a biblical image to guide environmental thought, it must be the New Jerusalem, the final culmination in a harmonious city where the groaning and travail of nature itself finally cease. As revelation makes clear, that unimaginable conclusion will come only at the end of time and by God's final action. In the meantime, we can only hope to move closer to a condition that will ultimately not be the result of our efforts. To press that image too exclusively, of course, might lead us into thinking that we may simply turn nature into anything we think good. But few people have such a one-sided view today. The best balanced religious vision incorporates all the elements of justice and compassion and joy that we find within us, and that seem largely absent in nature as we know it, with a well-informed scientific approach to problems that are largely practical. Some people may not find this sober path inspiring, but that may be precisely the problem. To expect our worldly existence to result in perfect liberation, this side of Paradise, is a delusion. Contrary to the emphasis most religious writers put on finding joy in nature, our current environmental task is more a matter of head than of heart.

Our heads, of course, always need reform. It is simply part of human existence to recognize not only sins like greed and selfishness but also innocent ignorance. As we have seen in previous chapters, for a long time we were unaware of what we were doing to nature, unaware and mostly innocent. And rightly so, since human activity had minimal impact. Today, we have plenty of impact. Anyone today who remains unaware of environmental issues cannot plead innocence. The other side of the coin, however, is that making the environment the radical ground of human existence in the way many environmentalists do is a snare and a delusion. In the same way that radical movements for liberation have misconstrued social justice to mean a dismantling of civilization and its material productivity because of injustices and failures, an environmentalism that seeks human and environmental liberation through a return to an imagined past or an illusory green future portends several forms of disasters for us and for nature.

The way we remain faithful to our past — to the hunters who learned about animals to track them; to the gatherers who identified unknown kinds of food; to the agriculturalists who developed new strains of wheat, maize, and potatoes; to the inventors who tamed fire, water, and earth for human purposes — is to carry forward their work in a new way. We will need to learn more about animals, plants, and minerals to preserve them, as far as possible, and ourselves. We will likely face many disappointments and triumphs, as did our ancient ancestors and forebears in every age. But we will have to try to make of the world something of which we now have only a dim inkling.

Civilization or Nature?

All of this is worlds way from the unfortunate state into which the environmental debate has fallen for some people. Our choice is not between a robust civilization and a mostly undisturbed nature. Our choice is between a better civilization and nature, on the one hand, and a fearful flight from our own powers that, given our numbers today, will spell disaster for civilization and nature. Had we in the West not, for example, substituted coal for wood and then developed oil and gas for fuels over the past centuries, the forests of the world would be in far worse shape than they already are in some places. The climate changes we now fear from burning fossil fuels do not negate the value of those adaptations; they merely call upon us to meet another set of challenges. Similarly, had we not learned how to produce more food on less land without beasts of burden and the cleared pastures they require for food, the needs of the vast populations we must now feed would have made our future even grimmer. Without the efficiency of modern industries and agriculture, without the much-hated international corporations themselves, solutions to our future challenges would be few and their availability to the billions who desperately need them slight. Entrepreneurs, engineers, and inventors will be the salvation of the earth in far more concrete ways than most environmentalists.

All this no doubt sounds hopelessly shallow to people who think that only a radical spiritual and political restructuring of the world will save the earth. But the burden of proof is on them. We have begun to get many environmental problems in hand. We are tackling others. Most people today — in developed and developing countries — believe that they can simultaneously have prosperity and a better relationship to the earth. Making both of those outcomes fit better with a renewed spirituality and appreciation for creation will be a worthy goal for religious people in the future. But to think that spirituality must be recovered at the price of indisputable human goods puts many religious environmentalists simply outside the conversation of mature, democratic people.

Much ink has been wasted debating whether economic growth or environmental value should take precedence generally or in specific instances. Both of these positions ignore a fundamental fact of modern societies. Wealth and well-being today stem far more from human innovation and scientific research than from the mere exploitation of so-called natural resources. It is entirely feasible today to expand the gross national product without harming the "gross earth product," and the way to do that lies in technological innovation. Not everything that technology can do is good. But the notion that natural resources per se are the basis of wealth is largely a belief of the past. Growth with environmental safeguards in the future means the growth of efficient, nondamaging processes. The sooner they exist throughout the whole world, the better for man and nature.

It is as a useful rule of thumb that when people's own interests are involved, you may have to take their position with a large grain of salt. Thus much of the outcry from industries over the years about the impossibility of dealing with environmental problems without unacceptable human costs has usually proved wrong; capitalism seems quite capable of finding ways to meet the dual demands that democratic politics throws up to save both people and nature, despite the resistance. At the opposite extreme, many environmentalists believe that greed and "industrial plunder," rather than an earlier ignorance about the environmental effects of industry, lie behind most environmental problems. To judge by the speed with which several corporations around the world have reacted to concerns about the environment, that explanation is at least in good part wrong.

Environmental questions are also beginning to be introduced into political judgments through the notion of "ecojustice." This neologism is merely meant to state an obvious fact: that poor people — whether in the developed or developing countries — live in areas less protected by environmental laws or even zoning codes and often bear a disproportionate share of environmental risk. Landfills, toxic waste sites, and industries are typically closer to, or actually in, poorer neighborhoods. Those people do not normally have the resources to fight environmental battles, so they often have to live with situations that others might find intolerable. Some religious denominations have opened up offices for ecojustice, and the Environmental Protection Agency is about to do the same.

A hardheaded look at these situations, however, reveals that some trade-offs are inevitable. On the one hand, no one wants to live with the potential health hazards and sheer lack of aesthetic appeal associated with polluted areas. On the other hand, some communities have little to offer other than their location and derive no little economic benefit from factories or other facilities that might not be permitted elsewhere. In a notorious 1997 case, residents of Convent, Louisiana, a mostly poor and minority community, lost the installation of a much-desired plastics plant that would have brought many jobs to the area. The plant would have had state-of-the-art pollution controls and would have exceeded all existing environmental standards. Unfortunately, a group of environmental-law activists blocked the project in the name of environmental justice. The vast majority of residents were willing to accept the plant's presence for the benefits it would bring. It was perhaps not an ideal choice, but it points us towards the inevitable conflicts that arise in decisions that need to be made about environmental questions. Sometimes people may choose for economic reasons to live with a less clean environment. Sometimes they will make the opposite choice. Justice in a democratic society consists not in attempting impossible equalization of environmental quality but in empowering people in these situations to get as fair a deal in economic and environmental terms as possible.

Wrong Emphases

A telling sign that something is wrong in the large-scale national and international environmental movement is that great attention is devoted to speculative or secondary problems, not only for such minority communities, but also in the international arena. Internationally, much attention is paid to issues such as global warming and species loss, very little to the concrete challenges and the outright death of poor people everywhere. Even when people in poor countries are noticed by environmental writers, quite often their poverty is blamed on the First World. In yet another twist on the living-in-natural-Eden theme we have seen in various religious environmentalists, the assumption behind much criticism of the First World is that the South and other poor areas once were in fine shape and still would be if the North and the European cultures had just stayed away. In this view, colonialism, extractive enterprises, or polluting factories spoiled an idyllic or, at the least, quite pleasant existence. The behavior of people who actually have a chance to experience real economic development or escape their current circumstances tells quite a different story.

As far back as the beginnings of the international environmental conferences in the 1970s, poor peoples — whether in developing or developed countries — have displayed an instinctive resistance to environmentalism. For them, it often appears to be just one more way for developed countries to ask others to pay the costs for problems already-industrialized societies have created. In 1971, a group of Third World countries, over the objections of the First World, declared that advanced technology "represents at the present stage the best possible solution for most of the environmental problems in the developing countries."[1] For nations with growing populations, that strategy seems the only alternative to allowing many people to die of starvation and natural environmental hazards. Many First World environmentalists fear that strategy will multiply what they regard as evils in new settings. They would like to see developing nations retain less developed ways presumed to have smaller environmental impacts. But the choice comes down to providing bread and health care for the poorest of the poor, on the one hand, and stopping such development for allegedly global reasons, on the other. As it turns out, the Third World countries were right in 1971. According to the 1994 UN *Human Development Report,* in the developing world crucial measures such as life expectancy, literacy, and nutrition have been steadily improving. Despite swift and continuing population growth, the percentage of people living in "abysmal" conditions declined from 70 per cent in 1960 to 32 per cent in 1994.[2] These figures indicate that we are still far from having universally solved the problem of development. But they are no small signs of hope for the world and, without the kind of development we have been promoting, might have reflected even worse horrors.

Some observers have noted that the West's concerns have elbowed aside the real needs of the poor. Gregg Easterbrook starkly declares: "What environmental problems kill human beings in number today? Not Alar or ozone depletion. What kills them is dung smoke and diarrhea." The 1992 Rio Summit, he says, is a case in point:

> Institutional environmentalism focuses on the real but comparatively minor problems of developed nations in part to support a worldview that Western material production is the root of ecological malevolence. The trough of such thinking was reached at the Earth Summit in Rio in 1992. There, having gotten the attention of the world and of its heads of state, what message did institutional environmentalism choose to proclaim? That global warming is a horror. To make Rio a fashionably correct event about Western guilt-tripping, the hypothetical prospect of global warming — a troubling but speculative concern that so far has harmed no one and may never harm anyone — was put above palpable, urgent loss of lives from Third World water and smoke pollution.[3]

After Rio, says Easterbrook, several Third World delegates were angry about the First World's stranglehold on the agenda. Some felt particularly outraged over the way environmental groups and the Rockefeller Foundation have discouraged high-yield agriculture in Africa because, with its use of chemical fertilizers and other advanced techniques, it is not "sustainable." Easterbrook contends that, given their less-than-ideal circumstances, developing countries need intensive agriculture to feed their people at relatively lower cost to the environment. They also need water-treatment facilities to combat diarrhea. And for severe air pollution, the remedies are "paved roads [to reduce air-borne dust particles], catalytic converters, hydroelectric dams, modern petroleum refining, advanced high-efficiency power plants — the sorts of technology green doctrine considers outrageous."[4]

Millions of children die each year in poor countries because of impure water (usually because untreated sewage runs directly into drinking water sources), something easily fixable if we had the will. Similarly, many die of lung problems related to poor air quality. Quite often this is presented as a pollution problem, as if it were related to industrialization. In some places in the Third World, Mexico City prominently among them, that is certainly the case. But as the *New York Times* reported during the 1997 Kyoto summit on greenhouse gas emissions, 2.75 million children died in 1996 from poor air, sometimes linked to the forest fires that blanketed Asia that year. But by the *Times*'s figures, over three-fourths of those deaths were related to smoke from cooking in the home over wood and dung fires. Even dirty coal-burning electric plants in India and China generate energy that provides as much as fifteen times the power per unit of pollution as dung and wood fires. Getting electricity through various means to people in these situations is crucial, as is transmitting other now-basic technologies that will make human lives better and their impact on the environment lighter, even in developing countries.

When it confronts these facts at all, religious environmentalism often falls into predictable warnings about the "seduction" of non-Western peoples away from their traditional ways by the consumerism of the industrialized world. This is quite a remarkable argument. Native peoples, otherwise thought of as virtuous, will be seduced merely because better tools and farm implements, more affordable clothing, better sources of food, new forms of energy, become available? Assuming, as the defenders of these ways of life do, that developing peoples are perfectly viable and mature in themselves, it is a very censorious and condescending view. Developing peoples are adults who can make up their own minds about the relative utility and value of better products and technologies. When they are given the chance, their decisions are quite clear: they want to join others in a better life. Fears of homogenization of the world notwithstanding, people in developing nations have their own ways of preserving religion, culture, family, and ritual while engaging the larger global market. Here as elsewhere there are trade-offs, but that is the case with all human lives everywhere. Cultures are not sacred. Only the sacred is sacred.

Many of the religious environmentalists we have looked at put great emphasis on changing attitudes towards creation and relatively

little on addressing concrete environmental issues. Their general belief seems to be that if we get our notions about the Virgin right, the
problems with the Dynamo will be corrected or simply disappear.
Given the doubtful environmental record of premodern peoples who
had religious cosmologies roughly similar to those advocated today,
that contention seems less than certain. Anyone today who wants to
be taken seriously in the environmental debates needs to look very
carefully at science, technology, and the way those human achievements are used or misused. Religious environmentalists in particular
should examine the plight of the poor. For the most part, they have
failed to shoulder that burden.

Liberation Theology and Ecology: Leonardo Boff

The main exceptions to the rule, not very fruitful ones, however,
are the liberation theologians. During the Cold War, liberation theology was sometimes regarded as a kind of latter-day Marxist movement. Today, it is often thought to have substituted a "green" approach for the old religious sympathy for various "red," or socialist,
critiques of capitalism, industrialization, and liberal democracy. In
strict terms, the charge is false, if only because Marxism per se does
not harmonize very well with the underlying anti-industrial tendencies of modern religious environmentalism.

Marx was an advocate of industrial development as a human good,
hence his critique of the "idiocy of village life." And he mocked those
who thought that the mere changing of philosophical and theological
opinion would lead to liberation. For instance, in *The German Ideology*
he writes:

> We shall, of course, not take the trouble to enlighten our wise phi-
> losophers by explaining to them that the "liberation" of "man" is not
> advanced a single step by reducing philosophy, theology, substance
> and all the trash to "self consciousness" and by liberating man from
> the domination of these phrases, which have never held him in
> thrall. Nor will we explain to them that it is only possible to achieve
> real liberation in the real world and by employing real means, that
> slavery cannot be abolished without the steam-engine and the mule
> and the spinning-jenny, serfdom cannot be abolished without im-
> proved agriculture, and that, in general, people cannot be liberated

as long as they are unable to obtain food and drink, housing and clothing in adequate quality and quantity. "Liberation" is a historical and not a mental act, and it is brought about by historical conditions, the [development] of industry, commerce, [agriculture], the [condition of intercourse].[5]

It would be difficult to imagine him as sympathetic toward a return to agricultural societies. Indeed, Marxist states clearly pursued industrialization with a blindness towards its ecological effects that is unparalleled elsewhere in the world.

But in the confused way that ideas often find concrete embodiment in human history, there is a rough truth in the charge that some religious environmentalism has a red cast beneath the green. Marx believed that the contradictions internal to capitalism would lead to crises in which the exploited workers and proletarians would, by sheer historical necessity, be forced to rise up and take control of the means of production. That historical thesis proved false: despite many problems and inequalities, workers in capitalist systems share more in the wealth and governance of their societies than their socialist counterparts and have proved to be a stabilizing force in capitalist systems. The neo-Marxist hope that people in the Third World could be used as a revolutionary force against the capitalist First World also failed. Countries like the "Asian tigers" that joined the global economic system prospered, while those that could not or would not seek development lagged behind.

Yet the use of the poor, the marginalized, and the underdeveloped has returned in some religious environmental thought as a new way to present an old critique. As the Brazilian Leonardo Boff, one of the leaders of liberation theology, has formulated the question, the old critique of capitalist development may be given a new dimension: "The logic that exploits classes and subjects peoples to the interests of a few rich and powerful countries is the same logic that devastates the Earth and plunders its wealth, showing no solidarity with the rest of humankind and future generations."[6] Boff was once a Franciscan priest, another Catholic cleric whose quarrels with the Vatican and irregular personal life eventually led to his departure from the priesthood. His analysis provides a good window onto a prominent element in religious environmentalism.

Like many religious environmentalists, Boff sees the world as a "vast community," in which harmony and peace and desire for interaction are the emerging guidelines for thought and behavior: "people feel the need for a new use of science and technology *with* nature, *on behalf* of nature, never *against* nature."[7] Our current situation "must have been a deep mistake, some grave error in cultures, religions, spiritual traditions, and in the pedagogical processes of socialization of humankind that failed."[8] He notices, only to reject, the claim that cleaner, more efficient technologies will solve the environmental crisis: "Is it not an illusion to think that the virus attacking us can be the principle by which we will be made well?" The doubtful medical analogy aside (viruses are often used to develop serums to treat infections), Boff concludes that "technology does not produce benefits for all societies but only for those that control scientific and technical production by exacting heavy tribute (royalties)."[9]

But Boff directs his heaviest fire against what he regards as a false notion of development that merely seeks to manufacture ever more and different products, exploiting the earth as much as possible. Even the 1987 UN Brundtland Report, *Our Common Future,* which first made the term "sustainable development" popular, is suspect to Boff because the concept "never gets away from its economic origins, namely, rising productivity, accumulation, and technological innovation."[10] Disregarding many of the concrete indications of improvements across a broad spectrum that we have noted earlier in this chapter, Boff rejects the notion that development will enable the billions of poor people in the Third World to live better with less impact on the earth. Rather, he claims, "poverty and environmental deterioration . . . are the result of precisely the kind of development that is being practiced." Boff sees the world situation as a strictly limited one in which some can get rich only because others get poor. Capitalism, in this view, is "especially against nature."[11]

Boff finds the word "development" misleading; development would include other human factors than profit. What even UN documents promote is *growth,* and that is by nature disruptive of ecosystems. In that respect, he says, socialism and capitalism are barely different. Socialism tries to distribute benefits to society as a whole rather than allowing private appropriation of goods as capitalism does, but the underlying paradigm of exploitation is similar. For Boff, human history has passed through three phases: an *era of the spirit,* reflected in the cosmologies of early peoples; an *era of the body,* our period, in which we believe ourselves masters of the material universe; and a coming *era of life,* in which we will see ourselves as interrelated with all things.

The Amazon Story

As a Brazilian, Boff regards the treatment and mistreatment of the Amazon as encapsulating the whole ecological story. What we now call the Amazon was hundreds of millions of years ago under water. But through a series of geological shifts, it became the lush tropical rain forest with which we are familiar. So explosive is nature there that a few acres contain more species than exist in all of Europe. Human interaction with the forest was great even before modern times. Boff regards as a myth the idea that Amerindians were *"genuinely natural beings,* representatives of the peoples of the virgin forest and hence in perfect harmony with nature."[12] He sees those people as much culturally shaped as we are, the difference being that they regarded themselves as part of the forest and the forest as in evolution with them.

He is also concerned to dispel two other myths: that the Amazon is perfectly in balance and is therefore the lungs of the world, and that the Amazon is the storehouse of the world. For Boff, it is necessary to see that the Amazon exists for itself. Its cycles of carbon dioxide consumption and oxygen generation are for itself, as is its diversity: "Research has shown that the 'forest lives off itself' and largely for itself."[13] He makes these points, it appears, so that the Amazon may be defended in its own right. In his view, the forest was largely intact until 1968, when a series of Brazilian military governments began to encourage the clear-cutting and burning of areas for commercial exploitation.

Boff details how several ill-advised attempts to use the Amazon for large-scale rubber production, logging, agriculture, industry, mining, and electrical generation foundered on simple-minded approaches to a powerful ecosystem. In most cases, the forest simply refused to cooperate and the plans ended in failure. Several indigenous tribes were also harmed in the process. From these experiences Boff draws a general conclusion: "The observations we have made about the Amazon — and that we could also make about the Pantanal in Mato Grosso and the Atlantic forest in Brazil — are overwhelming proof of how misguided it is to pursue development along the lines of modernity."[14] Boff does not believe that the Amazon should simply be kept as a preserve; rather, we need to learn to use the available nuts and herbs in a way that the rain forest itself dictates to us. Contemplating that truth should provide a paradigm for all economic activity and our vision of ecology around the world.

Liberation or Continued Bondage?

There is much, of course, in Boff's analysis that calls for thought, and some of his positions may come as quite a surprise to people who think of liberation theology and the ecological concern it has embraced as merely a Marxist front. But there are also many problems in this analysis, beginning with Boff's wide claims about the Amazon. Indeed, the use of the Amazon as a planetary ecological paradigm may be precisely the wrong place to begin. Boff takes as a model the most complex ecosystem on earth, and therefore his solutions seem far less compelling for simpler ecosystems. The rubber trees planted by the Ford Motor Company in Brazil failed miserably owing to local conditions, but rubber trees were transplanted spectacularly to other parts of the world. Similarly, the Amazonian hydroelectric projects may have been ill conceived, but there is no question that electricity and other developments that call for large capital investment and industrial plants are needed to improve the plight of the very poorest inhabitants of Brazil. Boff merely takes some spectacular failures as proof that the approach is thoroughly mistaken; yet we know so little about the Amazon that, for that specific case, the best policy is probably to leave it intact, as far as human well-being allows, while pursuing other avenues of development.

In addition, Boff shows great myopia in believing that the logic of industrial production is identical with the fiascos in his own country. For example, people long believed that industries would foolishly use up resources leading to massive shortages. Boff several times cites the Club of Rome's 1972 *Limits to Growth* as having sounded the first warning about the future. But that document got virtually everything wrong: it predicted that vast shortages in minerals, petroleum, and natural gas would occur within two decades. We are already almost a decade past the crisis point, and the minerals it identified as scarce — gold, mercury, tin, zinc, copper, lead — cost less than before. Known reserves of coal, petroleum, and natural gas are massive. (Indeed, in the 1990s, several of the main oil-producing nations were in economic straits because of a glut that depressed prices.) It was not only, however, that quantities were greater than *Limits to Growth* realized in 1972; improved drilling and exploration techniques expanded the range of recoverable reserves. Entrepreneurship found ways to use fewer resources more productively, sometimes spurred by fear of future shortages. More important, human invention never remains wedded to one resource forever. We once used wood for heat, whale oil and other animal products for light. The discovery of coal and oil spared our forests and made it possible to do without whale oil entirely. In that sense, the much hated oil and coal companies have actually preserved parts of nature while giving us products like gasoline, which in the United States costs less per gallon than milk.

Boff also weakens his case by — truthfully — describing the behavior of the Amerindians who inhabited the rain forest long before the arrival of white Europeans. As we have seen earlier, what we wrongly think of as pristine Amazonian rain forest is actually the result of the shaping of that region by human interaction. Not surprisingly, the Indians sought ways to maximize yields of the trees and fruits that they found useful. It is a non sequitur that they practiced a kind of agriculture that did not disturb the forest. We do not know what the "natural" state of the Amazon would have been without their intervention, or whether that would have been better or worse for the global ecosphere. The Indian populations were small and, from our point of view, their action was not detrimental. But that gives us little reason to think that their behavior was ecologically exemplary in any real sense. They might just as easily done things in pursuit of survival — and probably did, given the complexity of the forests — that caused extinctions of species and long-term damage.

Indeed, according to anthropologist Allen Johnson of the University of California, some of the Amazon peoples today are quite cruel:

> The Machiguenga [of Peru's Amazon basin] are usually kind to
> each other and to household pets . . . but men, women, and chil-
> dren torture captured animals for the amusement of onlookers,
> with both the torturer and the group laughing riotously at the piti-
> ful cries of the victim. A children's sport among the Yanomamo of
> Venezuela and Brazil is tying a lizard to a string anchored to a stick.
> They then happily chase it around and around, shooting arrows
> into it. The same behavior in Europe or America can bring a stiff
> fine or a jail sentence.[15]

Citing one bad cultural practice or even many does not invalidate the
notion that there is something to learn from a given culture. But examples like these cast doubt on the most common assumptions about
the harmonious interactions between indigenous peoples and the
rain forest.

In fact, there is probably no deep difference between the ways native peoples and modern technologies seek material improvements.
Boff and others, for example, would probably deplore genetic engi-
neering of new species, while applauding the ways native peoples developed new strains of manioc, corn, and other foodstuffs by selec-
tive breeding. Yet Native American procedures sometimes led to the
kinds of irreversible changes that are usually regarded as dangerous
because they make human interference in nature necessary forever.
Corn, for example, had been bred prior to the European arrival in the
New World to the point where it no longer could survive in the wild
and reproduce without human planting. Native Americans actually
had to sow kernels, much like modern farmers.[16]

At bottom, however, Boff's opposition of primitive to modern has
little to do with the nature of human activity that he thinks he finds in
the one or the other. Though he occasionally concedes that even his
radically new approach to nature will entail uncertain human advance
along the path of cosmic evolution, he has a religious (in the full sense
of the term) repugnance towards the imperfect, wobbling advances
made in recent centuries towards human well-being through industrial and capitalist development. This leads him into a hardened stance
against reformed uses of modern technology. He argues, for example:

> The most recent arrangements of the world order led by capital and
> under the regime of globalization and neoliberalism have brought

marvelous material progress. Leading-edge technologies produced by the third scientific revolution (computerization and communications) are being employed and are increasing production enormously. However, they dispense with human labor and hence the social effect is perverse: many workers and whole regions of the world are left out, since they are of little relevance for capital accumulation and are met by an attitude of cruel indifference.

You might expect him to believe that getting other peoples into this cleaner, virtually dematerialized technology would meet with his approval. But he believes that the global market system, not the age-old conditions of scarcity and population pressures, exact tributes tantamount to holocausts from people who "die before their time": "Recent data indicate that today globally integrated accumulation requires a Hiroshima and Nagasaki in human victims every two days."[17]

In the final analysis, this hardened stance betrays a kind of *ressentiment* against development that can only be called perverse. Boff succumbs to one of the besetting temptations of religious environmentalism, calling for a spiritual purity of means and ends that condemns to their current plight the very people about whom he claims to care. To advocate shunning or dismantling the current systems that support human life without a truly careful look at what specific measures will help peoples is irresponsible. Boff places great hope in the eventual development of world government through the United Nations and the establishment of a worldwide democracy that is also a "biocracy" and a "cosmocracy." But such a hope requires us, first, to believe that these new "-cracies" will arrive soon enough to help people currently threatened and, second, that the biosphere and the cosmos are far more harmonious than they actually are. In Boff's version of nature, there is rarely any struggle for survival, no muddling compromise with the less than perfect conditions of life on earth. Surprisingly for a liberation theologian who claims that the liberation of the poor means grappling with the concrete circumstances of their lives, he is finally more concerned with liberating people from what he takes to be mistaken First World philosophies and theologies than in offering them practical relief. He would have helped the poor more had he been a more faithful follower of the Marx of *The German Ideology*.

In a Secular Mode: Martin W. Lewis

Writing in a secular vein, Martin W. Lewis has rebutted such contentions in his *Green Delusions: An Environmentalist Critique of Radical Environmentalism*. As a former environmental radical, Lewis is familiar with all the usual arguments for recovering a sense of awe at nature and connecting individuals and whole societies more closely to natural rhythms as a prelude to better environmental behavior. Lewis doubts that this will be helpful in merely practical terms. He dismisses as largely myth the claims that peoples "in tune" with nature are better behaved environmentally. Traditional ways of living on the land take a heavy toll on nature, especially where populations are large. Instead of the Arcadian environmentalism, based on Rousseau, that permeates much environmental thought, he proposes a Promethean environmentalism that "embraces the wildly creative, if at times wildly destructive, course of human ascent."[18]

Despite current problems and outright crises, he believes that we would do better to "de-couple" ourselves from nature still more. De-coupling for him does not mean that we should ride roughshod over natural processes. Rather, by concentrating people in urban and suburban clusters, pursuing new "nanotechnologies" that will carry out at the atomic or molecular level work that now requires large machinery, and reforming older technologies to eliminate pollution, we have the best chance available to us to be kinder to human beings and to nature by leaving her more alone.

Lewis finds a dangerous extremism in the green movement's covert, and sometimes overt, argument that civilization itself is the problem. "Reform" environmentalism, for the green movement, gives the illusion of improvement in the environmental situation while only delaying the radical destruction and reconstruction that must be done. Even the green embrace of "appropriate technology" hides a near religious faith that it would really be best for man and nature if there were no technology at all. "Appropriate technology . . . often turns out to mean little more than well-engineered medieval apparatuses. . . . [T]he systematic dismantling of large economic organizations in favor of small ones would likely result in a substantial increase in pollution, since few small-scale firms are able to devise, or afford, adequate pollution abatement equipment."[19] Those who advocate a return to "natural" energy sources like wood instead of petroleum and the dreaded nuclear option are oblivious to the far greater pollution caused by wood burning and its potential to reduce the United States to "a deforested, soot-choked wasteland within a few decades."[20]

To those who simplemindedly think small is beautiful and big is ugly, Lewis introduces a useful set of discriminations:

> "Primal" economies have rarely been as harmonized with nature as they are depicted; many have actually been highly destructive. Similarly, decentralized, small-scale political structures can be just as violent and ecologically wasteful as large-scale centralized ones. Small is sometimes ugly, and big is occasionally beautiful. Technological advance, for its part, is clearly necessary if we are to develop less harmful ways of life and if we are to progress as a human community. And finally, capitalism, despite its social flaws presents the only economic system resilient and efficient enough to see the development of a more benign human presence on the earth.[21]

In addition, he doubts whether it is true that economic growth is inherently unsustainable and whether it is wise to ask Third World countries to shun industrialization and "isolate themselves from the global economic system."

Almost every one of these points contradicts common environmental assumptions. But, more telling, Lewis thinks that such notions persist out of ignorance: "the foundations of green extremism are constructed upon erroneous ideas fabricated from questionable scholarship. Radical environmentalism's ecology is outdated and distorted, its anthropology stems from naive enthusiasms of the late 1960s and early 1970s, and its geography reflects ideas that were discredited sixty years ago. Moreover, most eco-radicals show an unfortunate ignorance of history and a willful dismissal of economics." Perhaps even more disturbing is the radical assumption that "the roots of the ecological crisis lie ultimately in *ideas* about nature and humanity."[22] He adds later: "Advances in solar power will not come about through holistic inquiries into the meaning of nature."[23]

It is sometimes bracing to encounter an apostate from a faith or a convert to a new one. Lewis does not disappoint on either count. He knows from the inside the way that environmental extremism has insinuated itself into the nooks and crannies of American and global thinking. It remains in its purest form a decidedly minority position, but its weakness in terms of sheer numbers is made up for by its influence on the commanding heights of the culture in universities, political debates, and elsewhere. In particular, he attacks the notion that "economic growth is by definition unsustainable, based on a denial of the resource limitations of a finite globe." He contends that stopping industrialization in the Third World will guarantee high rates of population growth. Much depends, instead, on how new development occurs: "While the global economy certainly cannot grow indefinitely in *volume,* it *can* continue to expand in *value* by producing better goods and services ever more efficiently. . . . If we fail it will be in devoting too few of our resources to technology, not too many. . . . Technologies, not natural resources, provide the essential motor of economic development."[24]

Among the explanations for Third World poverty, Lewis finds a common neo-Marxist thread. Barry Commoner has argued that the poorer nations were impoverished through Western colonialism. Lewis counters that Afghanistan, Nepal, and Ethiopia were barely touched by colonialism and are hardly models of non-Western, indigenous development. Colonizers did harm some indigenous economies, but Lewis believes that, on balance, Western colonialism is overemphasized in radical analysis. Nor does he think much of the argument that current difficulties can be solved by global redistribution. Jeremy Rifkin, for instance, comments that "as long as we in the United States continue to consume one-third of the world's resources annually, the Third World can never rise to even a semblance of a standard of living that can adequately support human life with dignity."[25] But Lewis sees this as mere wishful thinking given the extreme unlikelihood that America will voluntarily reduce living standards. Other analysts believe, as we saw in the case of Leonardo Boff, that the developing nations are dependent on the developed and are therefore held in thrall. Lewis observes, "Many proponents of dependency theory even argue that famines are largely a by-product of global capitalist economics, implying that hunger is virtually unknown in precapitalist and socialist societies."[26] While dependency theory withered in secular circles as exceptions exploded the notion, most environmentalists, says Lewis, are unaware of the shift in social analysis.

Lewis believes that only robust and ecologically sound development can remedy both population and land pressures in developing countries. "No program committed to small-scale technology and economic autarky can ever foster genuine development."[27] China's rural iron production, for example, was a disaster. He advocates sustainable development and believes that fast growth is possible if it basically respects ecosystems and efficiently reduces the amount of energy and raw materials used for each product. Though some small-scale projects may fit certain circumstances, Promethean environmentalism recognizes the need for industrialization and urbanization of the right kind. These latter processes are usually opposed by environmentalists for the manifest problems they bring, but Lewis notes that there is no better way to reduce pressure on the land than to bring people into cities. In the radical view, figures like Lewis and mainstream environmental organizations like the Sierra Club and the Audubon Society have "sold out to the despoilers" because they are willing to compromise with various interests for incremental gains. Lewis remains a committed critic of the current situation, but he worries that environmental radicalism "in seeking to dismantle modern civilization . . . has the potential to destroy the very foundations on which a new and ecologically sane economic order must be built."[28]

Al Gore and Reform Environmentalism

Boff is right, however, that we face three basic options: the old industrial course, reformist ecology, and a radical new combination of ecological and human liberation. No one of any intelligence believes any longer in the old industrial course, so the choice is basically between a more or less reformist and a more or less radical restructuring of human action. But the line between these two is not as clear as Boff's formulation suggests. He himself wavers between specific reform elements and a much vaguer call to radicalism. The same can be said of other prominent figures, among them America's most visible environmental leader, Al Gore.

Gore is a particularly interesting case because he clearly needs to preserve his political viability even as he calls for radical change. Some of his critics see a cynical personal calculation in his attempt to reconcile these two elements, but it might be fairer on the face of things to say that Gore's predicament demonstrates the kinds of challenges that will confront any leader who really hopes to move opinion and practice in the developed world. In addition, Gore makes religious claims about his environmental position. The several tensions between his political aspirations, environmental analysis, and spiritual commitments reveal a lot about the current debate.

It is curious that Gore's political action assumes that what most needs addressing is widespread belief in the rightness of the old technological path. In his highly publicized book *Earth in the Balance* he asserts that we must make "rescue of the environment the central organizing principle for civilization," because we confront "the current logic of world civilization," and "we must all become partners in a bold effort to change the very foundation of civilization."[29] Environmental issues at times seem to be for him not one among several principles, like justice, liberty, freedom of conscience, and human inventiveness, but *the* central organizing principle, a position in potential tension with our constitutional order.

Gore cannot follow through on this, however, because it would require far more radical changes than he lets on in his public speeches. In his book, he is far more open. We need, he says,

> an all-out effort to use every policy and program, every law and institution, every treaty and alliance, every tactic and strategy, every plan and course of action. . . . Minor shifts in policy, marginal adjustments in ongoing programs. Moderate improvements in laws and regulations, rhetoric offered in lieu of genuine change—these are all forms of appeasement, designed to satisfy the public's desire to believe that sacrifice, struggle, and wrenching transformations of society will not be necessary.[30]

That has the old Baptist hellfire-and-brimstone air and is the covert creed of much religious environmentalism. Gore also speaks of coming "Holocausts" and "environmental *Kristallnacht*"—as if the human impact on earth were a sinister, deliberate project rather than an unintended consequence.

Gore has hectored the American public about the need to undertake a massive commitment akin to fighting the Nazis and reconstructing Europe. In this scenario, he plays Churchill and Marshall, his opponents are Chamberlain and the isolationists. But for a politically shrewd man, he may have badly misgauged what it will take to get people to make the kind of sacrifice he envisions. Modern democracies tend to be moderate and pragmatic regimes. It took a long time for England and an even longer time for America to get involved in the struggle against Hitler. There was much wishful thinking and self-delusion in delaying responses to his intentions and actual action. Democracies go to war — or resort to wartime kinds of measures — only when threats become palpable. So far, no environmental threat has reached that serious a level. Some, like global warming, may yet prove their potency. But for the time being, asking people to make a "wrenching sacrifice" for a threat that may prove to be an illusion, as no small number of past environmental scares have been, is to ask for what democratic peoples by nature will not do.

Unlike almost all religious environmentalists, Gore and his team truly know some things about science, including the geological record. They examine some of the natural phenomena that we looked at in chapter 2, such as the effects of volcanoes and natural fluctuations in climate, but give them a different twist. For them, volcanic eruptions do not show some of the instabilities of nature and the earth's natural ability to recover. Rather, they show the fragility of climate. Gore seems unaware that natural interactions in the biosphere result in anything other than balance. For him, natural fluctuations in global temperatures, for example, are only long term and gradual. Writing before the publication of Maureen Raymo's work showing that natural changes can be large, sharp, and rapid, he comes to the conclusion that what we are doing at present is unprecedented.

He is right that "until our lifetime, it was always safe to assume that nothing we did or could do would have any lasting effect on the global environment."[31] But this does not lead him to exonerate past human activity on the grounds of ignorance. Like many of the figures we have already seen, he locates the whole problem of the environment in the West's Cartesian separation of mind and body. He accepts the Gimbutas theory of millennia-long goddess religions, about which he admits we do not know much but then claims that "its best documented tenet seems to have been a reverence for the sacredness of the earth....."[32] Elsewhere, he makes the remarkable leap that "but for the separation of science and religion, we might not be pumping so much gaseous chemical waste into the atmosphere and threatening the destruction of the earth's climate balance."[33]

In short, Gore's position is an odd mix of relatively well-informed scientific thinking and the usual over-generalization about remedying the environmental problem through reform of our thinking about creation. Monotheism, he concedes, was empowering until the modern developments in the seventeenth century, though there was some problem even as far back as Plato and his admirer, Augustine. Aristotle and Aquinas were closer to the good integration of body and soul. But in recent centuries, "we have chosen to leave the garden," and "we are, in effect, bulldozing the Garden of Eden." One of the commandments of our time is, "Thou shalt preserve biodiversity." And the concluding question is: "How can we glorify the Creator while heaping contempt on the creation?"[34]

A Dysfunctional Civilization?

Though Gore reaches for religious language at solemn moments, the overarching image through which he frames the environmental debate is not sin and redemption but the already suspect language of pop psychology. He devotes a whole chapter to denouncing our "dysfunctional civilization." In a Christian view, of course, all civilizations are dysfunctional in one respect or another because all are composed of fallen, and, in a sense, therefore dysfunctional human beings. But it strains belief to speak of America as filled with "enablers" who are "in denial" and refuse to "face their pain."[35] "We don't really live in our lives," Gore remarks cryptically at one point, presumably because of a deep disconnection between our psyche and the material world.[36] Our fumbling attempts to set a new course are a kind of "mid-life crisis" for the species.[37]

Is any of this even remotely true? And do people really care so little about the environment? Every major newspaper carries news of environmental threats almost daily. Television and radio regularly follow whatever environmental story happens to be most prominent. Students from kindergarten through graduate school are introduced to ecological issues. Every survey of the American people shows that about two-thirds of us place environmental concerns high among the issues we consider most important, right up there with saving Social

Security and even above reducing taxes. We just don't go as far a Gore would like. Among developed nations, America spends the highest percentage of GNP on the environment, though you would never guess this from what is said in international forums. All this may not run very deep or move the political and economic levers of power, but public expression of this sort is hardly evidence that talking about serious environmental issues is taboo.

Moreover, Gore takes the old psychoanalytic line that resistance to his interpretation of the direness of our situation proves its truth. A less Freudian interpretation might find that perhaps the resistance to panic over the environment may be the result of a realistic appraisal of our situation: we basically are on the right track but need further work. But to concede this would be to fall openly into a reformist camp that Gore studiously avoids.

For people shaped by the Bible, the psychoanalytic approach, however palatable in current political circumstances, errs not in going too far but in not going far enough. Suppose, as a thought experiment, that the government instituted compulsory counseling to liberate us from our alleged environmental numbness. Suppose, even, that the counseling worked. It might open many peoples' eyes to things that they had never thought about. But then what? If the biblical tradition has anything to say to us, it is that individuals and whole societies would still be subject to sin—pride, greed, foolishness, selfishness — to say nothing of simple human limitations. In any fair reading of history, we got ourselves into environmental problems largely through ignorance and a misplaced confidence in our growing powers. The most urgent requirement for the renewal of Christian thought about the environment is humility about nature and about human nature.

The parts of Gore's analysis that have gotten the most attention, however, are his specific proposals. The centerpiece of his program is what he calls a Global Marshall Plan, under U.S. leadership. Gore has been one of the promoters, of course, of the large United Nations efforts on the environment but, whether from conviction or American political expediency, in his book he is skeptical of the UN. Third World governments bear the brunt of much of the environmental blame in their own countries because of the corruption, tyranny, and ineffectiveness they often display, says Gore. Instead, he sees America as the only likely global leader that will combine "social justice, democratic government, and free-market economics." Perhaps in an effort to gain support from American conservative elements, he also adapts the appeal of the Reagan-era Strategic Defense Initiative and calls for us to create a Strategic Environmental Initiative.

As welcome as some of these positions are—especially U.S. leadership rather than the whims of international bodies—finally, Gore's stand is ambivalent. On the one hand, he produces lists of technological and political measures that should be implemented. We need: an Environmental Security Trust Fund, a Virgin Materials Fee, substitution of new and better technologies for older ones, high mileage requirements for cars and trucks, universal energy efficiency and conservation, tree-planting programs, and the phase-out of ozone-destroying chemicals. We ought to set up a Digital Earth data bank that will help us track climate changes. And the list goes on and on. The question is whether commitment to "social justice, democratic government, and free-market economics"—textbook articles of faith for any American politician — is as easily reconcilable with Gore's wide-ranging proposals as he thinks.

Gore has criticized the real radicals of Deep Ecology and Earth First! for their anti-human stance and outright callousness about human life. He regards blaming property ownership, capitalism, and democracy for environmental problems in the same light as seeing slavery as intrinsic to those same institutions. Environmental degradation is, like slavery, a contradiction of our deepest principles and must be cast out from good human institutions. Whether this pro-American declaration fits with Gore's environmental program, however, remains problematic. Perhaps because of certain alliances in the environmental movement, he is wary of ever presenting himself as a mere reformist. Technology alone, he often warns, is not enough, though the technological recommendations, backed by political incentives and penalties, must clearly be the centerpiece of a Global Marshall Plan or Strategic Environmental Initiative. Indeed, he warns, "as we focus our attention more and more on using technological processes to meet our needs, we numb the ability to feel our connection to the natural world."[38] But it is difficult to see how other than in an emotional or "spiritual" sense, this numbing matters to the Global Marshall Plan.

Another "Liberation Theology": Thomas Derr

In his 1971 *Ecology and Human Liberation: A Theological Critique of the Use and Abuse of Our Birthright,* the distinguished theologian Thomas Derr began a twenty-five-year critique diametrically opposed to many of the theological views we have been examining here. Derr asserts: "I wonder in the end whether the desacralization and historicization of nature are not better guides to good environmental management."[39] In Derr's view, the growing attempt to include nature in the great commands to love God and neighbor is an alien and potentially harmful addition to biblical thought. His precise reason for doubting the value of making nature the third of these commandments is that "love and reverence are due nature only derivatively, as the creation of a good God, and that nature may not be personified, endowed with selfhood, or otherwise treated on a par with man in Biblical religion."[40] The more usual environmental approach by religious people is that nature has intrinsic value, perhaps in some sense even rights. And religious environmentalism assumes that the doctrine than man "is set apart from the rest of creation is made to issue automatically in arrogance towards, and exploitation of nature; whereas in reality . . . the Bible will countenance no such consequence." Derr even takes a strong stand against those who point to the language in Revelation 21:1 about a "new heaven and a new earth" as meaning that the earth possesses a kind of quality that can be redeemed. We are conceived alongside a surrounding world, but that does not mean that nature is as valuable without us as with us. Nature is God's agent, but it is not free: "It should therefore not be considered as having a career with God parallel to man's, but a place in salvation history as a vital yet subordinate element to man."[41]

He also roundly contends that those who would "remystify" the world would harm our thought rather than helping. To begin with, he observes, "Nature scarcely exists unaltered by man, and where it does it proves a very ambiguous and unreliable guide for human conduct. As man is part of the natural order, it would be difficult in any case to see why the abstraction nature-without-man should be taken as normative. . . . Furthermore, an uncritical reverence for unaltered nature would oblige acceptance of disease and passive acquiescence in the face of other natural disasters like floods, earthquakes, and forest fires."[42] For these and other reasons, Derr judges that the remystification of nature "contributes to ethical irresponsibility."

He notes that the process philosophers and theologians, followers of the great twentieth-century mathematician and philosopher Alfred North Whitehead, have tried to grapple with the manifest evils, from our perspective, that exist in the world. Their solution is to propose a loving but limited God, involved in the process of trying to bring greater good out of the cosmos by "luring" free beings towards love. The world, in this view, is not so much God's creation—since it escapes his control—as a process towards some more loving future. God cannot be omnipotent or he would not allow the wholesale slaughter of beings by one another or their demise from natural forces and diseases. For Derr, this line of thought is a vigorous and philosophically valuable enterprise but incompatible with any faithful reading of the Bible. The mere fact that some theologians have to resort to positing a God of limited powers because of what they see in nature further alerts us to the difficulty of putting nature on a par with love for our human neighbor and God.

It is the paradox of human existence that—despite all the imperfections and threats in nature—we feel qualms, guilt, shame over our need to use other creatures to our benefit, when those same creatures feel nothing of that sort towards one another or towards us. Yet a kind of humble anthropocentrism remains in the heart of some of the most committed religious environmentalists. Derr observes that, after a trip to India, Lynn White, Jr. (the principal proponent of the notion that by desacralizing nature and giving man dominion, biblical religion led to perverse anthropocentrism in the West) expressed a desire to grind up the sacred cows for hamburger to feed the poor. For Derr, the desacralization of nature is not a vice of the West but a virtue, so long as it is rightly and intelligently understood. He seems to agree with Martin Lewis that de-coupling man and nature can lead to benefits for both. Indeed, if we think of the way that, say, Al Gore tacks on the need for spiritual awe and reverence at the end of a long list of hardheaded technological fixes, it would be difficult to say what natural sacrality brings to concrete decisions, other than a greater enthusiasm to avoid needless destruction and to manage prudently parts of nature that might otherwise return to harm us.

In addition to the value Derr finds in desacralization, he also sees the historicization of nature — in which man and nature evolve together — as having merit:

> To think of nature only as an unchanging condition may have certain uses, as in the case of the famous image of "spaceship earth," but it must not give too static a picture, eliminating the possibility of responsible transformation of nature in accordance with human needs. Nature and man evolve together. The future is open for both. The pattern is linear, not cyclical. The Biblical God is the God of justice, and it is no coincidence that He acts and is revealed historically. Nature gods like the Canaanite Baalim are fertility gods, concerned with the annual agricultural cycles, not with justice. They are institutionalized in socially static patterns. They support the status quo. The Biblical God is the God of the open future.... Nature, His gift and His instrument, is made by Him to serve His historical purposes.[43]

These are the claims, of course, that much religious environmentalism would wish to escape. They sound very much like the kind of ideology that such believers identify with the anthropocentric and "theistic" domination of the earth.

But Derr denies that this religious notion of coevolution should be read as a counterpart to the old Enlightenment ideology of progress. The Bible has a linear story, but it also warns of dark events, apocalypses, and judgments to come. Orthodox theology knows that the wheat and the tares grow simultaneously. It was a kind of Christian heresy, and a utopian one at that, that led to the secularized notion of progress. Derr rightly adds: "The mere passage of time accomplishes nothing of an automatically moral character." So even as Augustine's use of Christ as the "straight way" and the Bible's concern for achieving justice have given birth to a this-worldly progress, the real progress sought by the Bible does not disdain material, economic, political, and social improvements—but expects even more.

The Bible's desacralization of nature and lifting up of time, Derr believes, have advantages for us and nature. He is harsh on those who think natural harmony is somehow preferable to or more spiritually justifiable than human efforts to secure certain outcomes. In particular, he finds wanting the claim that we should abandon the linear rationality of the Bible for nature's cyclical ways, "To assert that 'We need to remakes the earth in a way that converts our minds to nature's logic of ecological harmony' is to gloss over the way that 'harmony' is achieved: the big fish eat the little ones."[44]

Perhaps in the end, Derr serves as a useful reminder of one of the things biblical principles have always warned about: idolatry towards nature. The kind of liberation from guilt over using other beings that we seek cannot come from according nature a status that it does not intrinsically possess. We can only seek to live worthily the lives we preserve at what we regard as cost to others. Furthermore, our human quest for justice towards one another demands human qualities that we do not find in nature. The assumption on the part of many religious environmentalists that our injustice towards one another is the paradigm for our injustice to the natural world (or vice versa, depending on the figure) may really confuse two things to the detriment of both. The "Joint Appeal by Religion and Science for the Environment," which counted Carl Sagan, Al Gore, and E. O. Wilson among its signers, asserted: "We affirm in the strongest possible terms, the indivisibility of social justice and the preservation of the environment."[45] This is a weighty claim often repeated in religious literature on the environment. But it seems to hope that we may readily combine two things that we desire, each quite difficult in itself, into some higher unity. We all desire human justice. Associating it with the environment may seem to strengthen the case for nature. But Derr may be closer to the mark when he says, "The truth is that sound ecology and the practice of justice are different concepts."[46]

CONCLUSION

"What Did You Go Out Into the Wilderness to See?"

On a hill near my home in Northern Virginia lie the remnants of the winter campgrounds of the Native American tribe known as the Doegs. It is easy to see why they chose the site. Large hills all around funnel water into a stream and floodplain there, making it a natural gathering place for wildlife. Deer still come quite often to drink; they are by now so used to the human beings who walk the paths through the woods that does and fawns are not particularly fearful until you get very near. Waterfowl and other game were also once plentiful. Unfortunately, they no longer exist there. Insects are a bother, as they must have been for the Indians. The campground is on top of one of the lower hills overlooking the water, probably for the fresher air as well as for an unhindered view of approaching enemies. Sometimes in the evening, when the sun is slipping through the trees and there are no sounds of civilization, it is not difficult to imagine—perhaps accurately, perhaps not—a kind of peacefulness and simplicity that once may have made up human life in these hills and valleys.

We once had a single-lane wooden bridge across the stream. But because of the traffic that chokes Washington, D.C., and its suburbs, it was replaced a few years ago by a modern multilane, multi-million-dollar road. The old wooden structure got us over the water slowly but surely. The new concrete road is much faster, but it floods over and has to be closed whenever there is heavy rain. Whether the engineers deliberately planned it that way for some purpose, practical or

241

environmental, I cannot say. The only thing certain is that even in the small corner of the world that I know in the most immediate sense, man and nature are still torn between two modes of coexistence. One seems to unite us into something larger than ourselves; the other seems to set us in greater or lesser opposition.

When I walk in the woods, especially with my younger daughter, I sense a reconnection with an origin, a living tie to a world of varied textures and modes of being, more varied in some ways than the houses with computers and cable television that dot the hillsides. Nature in these moments is an "antidote for civilization," as an upscale travel club bills its Caribbean vacations. And that is part of the problem. For lurking within the quite proper sense of ease and sublimity many of us feel in woodlands — the feeling that gave rise to Longfellow's forest primeval and Tolkien's Middle Earth — is the hope that we can escape the bewildering complexity of the human. That is a fantasy only certain people prosperous to the point of having forgotten about the human struggle with nature can allow themselves. As Santayana observed at the beginning of this century, "Nothing is farther from the common people than the corrupt desire to be primitive."

For primitivism, even in small bites, also reminds us of other human longings and fears. Sometimes when I am walking near nightfall in the woods and the bats take over from the birds in keeping the insect population down, a sense of the more threatening and inhuman side of nature comes over me. Then the woods seem not so much an origin enriched with literary associations as a trackless confusion, more like Dante's disorienting dark wood than an uplifting fabric of fellow creatures. At such moments, it is easy to see how the consciousness of another world beyond nature may have grown with the development of language and the creation of safe human settlements. But it is difficult to believe, as some have theorized, that this rise of human culture departs from our better instincts. Built into some of the deepest recesses of the human psyche are instincts about the dangerousness of the world. Few human beings sleep with arms or hands spread out and exposed — they might easily be vulnerable to attacks. Similarly, it appears that the development of human sight was one of the physiological advantages we achieved over animals that sense more by hearing and smell, hence our preference for open spaces where we can see potential threats coming rather than dark jungles where we may be easily outmatched by wild beasts.

Human beings have an instinctive fear of the dark. Ghosts may be just one manifestation of the archaic, unconscious dimensions of the psyche that make us concerned about what lurks where we cannot see. The dark hides from our keenest sense organ the vital information and advantage we need, given our relative physical weakness compared with predators. The image of a tribe around a campfire is a soothing one. Fire and light help ward off the beasts that we may more easily contend with during the day. A camp, however, is reassuring not only because it shows human warmth and community, but because it marks out a human space different from unrelieved natural challenges.

Even the relatively secure green zones created during the development of suburbs outside the capital of the most powerful and advanced nation in the world still bear witness to the double face of nature.

When Jesus asks the first time in Matthew 11 what people went out into the wilderness to see, he answers his own question with another: "A reed swayed by the wind?" What are we to make of this? Is Jesus, the central figure of Christianity, displaying a characteristically Western blindness about the value of nonhuman nature, in this case with a slighting reference to what we think are environmentally valuable wetlands? Or what shall we say about his next answer to the same question: Did you go out to see "someone dressed in fine clothing? Those who wear fine clothing are in royal palaces." Perhaps Jesus is saying here that the comforts produced by human artifice are hollow, a distraction usually connected with unjust social and political hierarchies.

As if these apparent contradictions were not enough, Jesus tells his listeners what they really went out to the wilderness to see: a prophet, "yes, I tell you, and more than a prophet." John the Baptist, we are instructed, is the greatest of all men born of woman, but even "the least in the kingdom of heaven is greater than he." John was Elijah, making the way straight for the one who was to come. But the generation that he and Jesus found living in their day neither danced when the two of them played a happy tune nor mourned when they played the funeral dirge. That is, the asceticism of John was unpalatable to them, as were the easier ways of eating and drinking that Jesus' disciples practiced. Why? Difficult as these passages are to interpret, we might say that the sin common to those who opposed asceticism and those who opposed abundance is that, at all costs, they were determined not to hear the good news, not to repent or to change their lives.

It would be a fanciful reading of these biblical passages (eisegesis, as the Bible scholars call it) to try to enlist them in the environmental debate. But it may not be so unrealistic to see here at least some biblical wisdom that might inform our thinking. We know from many biblical verses, if not from our own personal experience, that we have to be wary about attachments to fine garments, wealth, power, and influence. All of them pass, either because of changes during our lives or because we die. To put our hearts' treasure there is foolish. We also know from the Bible, and now increasingly from science, that the earth too is passing away, that the seas and land masses, the plains and what the Bible sometimes calls the "everlasting hills," are not everlasting. They, too, will pass, if not in our lifetimes then sooner or later, through various forces, natural and man-made.

Neither of these truths is particularly pleasant to contemplate, hence the resistance of some Israelites to John and Jesus. Whether we are ascetics or more easygoing in the robust Hebrew fashion, either we find our ultimate safety and salvation in faith in God, a frightening trust that demands everything, or we trust that created things, in nature or in human action, will provide us with peace of spirit until we go the way of all flesh. The latter course, though ultimately baseless, seems easier: we can see and touch, taste and feel the material world. By contrast, the call for a change of heart in John and Jesus sounds like a request that we turn literally to nothing: no thing, no activity.

As I have argued throughout this book, both unspoiled nature and human action may be viewed as participating in God's intentions for the world. Sometimes nature seems to manifest evils, sometimes our activities — even those undertaken to secure moral goods — reveal a sinister cast. We are enmeshed everywhere in mysteries, mysteries about why the world is as it is, mysteries about the self-deceptions and occasionally unexpected graces in the human heart. We have no clear-cut rules when we consider ourselves and nature. God seems to have built into the world some limits and an equally important plasticity that complicates our notion of what is natural. Human nature, too, is both limited and plastic, and the notion of what is human has expanded as all parts of the world have been brought into contact with one another in recent centuries.

Ultimately we are left with two truths. The first is that in creating the world, God had something in mind and saw that it was good. The other, the truth that is meant to set us free, tells us that beyond all the uncertainties and transience of this life, something, or someone, remains, that is the final resting place for human hearts. Or as Augustine expresses it at the beginning of his *Confessions,* "you have made us for yourself, and our hearts are restless until they rest in you." The peace we feel at times in nature is, properly understood, an anticipation of that rest. But if it is to be more than a mood, it has to find a deeper confirmation.

For many environmentalists, religious environmentalists among them, putting our ultimate faith and trust in something beyond nature is to risk reproducing the otherworldly neglect of nature that allegedly has been the besetting sin of the West since Plato. It seems to make the world a secondary realm unworthy of our full allegiance. Rousseau made a similar complaint against Christianity when he claimed that Christians, ultimately concerned about their eternal destiny, cannot be full citizens: they will look on the realm of politics as of only moderate importance. The criticisms of Plato and the complaints of Rousseau, though ultimately wrong in theory, perhaps point to some real problems in our practice.

If the truth that both we and the world are merely transitory leads us to give up striving for all the goodness and justice that we can achieve, we are poor interpreters of the Bible. The God who sends his only son into the world to die for it is not a God who takes a light view of history and human existence. But the certain knowledge that our lives and all of nature are destined to disappear may actually help us do better at our worldly tasks. We now think, on good secular as well as religious grounds, that making politics an ultimate good, in the fashion of certain ideologies of the past two centuries, leads to narrowing horizons and dangerous tyranny. As Dr. Johnson wisely observed: "How small of all that human hearts endure/ The part that kings or laws can cause or cure."

That kind of wisdom may also be of use when we turn to the environment. The development of human powers has led to a remarkable reduction of some ills and the outright elimination of others. Yet we know as well as we know anything that our technologies and therapeutic techniques cannot make human life in this world a paradise. To believe so would be an idolatry. By the same token, our concern for nature has to acknowledge the imperfections of the world even as it recalls the biblical assurance that creation is good. We may hope to manage our impact on the world with some intelligence, but recognizing our own limits and limitations vis-à-vis nature may help us ultimately to remain calmer and therefore to respond better in the face of environmental threats, whether natural or man-made.

The basis for value judgments in most environmental issues is natural versus artificial, the first, it is assumed, always good, the second, usually bad. This is only a symptom, however, of the skepticism in the last few centuries over speaking about what is good and bad per se. If we can claim that "nature" requires something, we don't have to engage in the acrimonious human debates about ethics that we feel are hard to resolve. We can claim absolutes for nature that we would not dare claim for ourselves. But anyone who begins to look at the constantly shifting, complex, discordant, and sometimes antithetical directions nature takes will be tempted to give up entirely on making natural claims. Animals and plants harm one another without our intervention, sometimes rendering one or another species extinct. "Bad things happen to good creatures," even in the absence of man, as one environmentalist has noted. So we may be forgiven — if there is indeed anything to forgive — if we do what the human race has done throughout history: decide what is good for human beings and the things they most value and try to fit those as well as possible into a biosphere that, properly managed, we may hope to preserve and enhance even as we know it will both nurture and threaten us.

At the end of all our speculation, we have to confess that we do not know why God made the world as he did, or why he chose to make a world at all. The best reflection on the biblical revelation has maintained that the world is not some realm that God rules over like the ultimate tyrant, but his communication of existence and love to beings that, strictly speaking, do not need to exist. God would have been perfect God without us or the entire universe. Yet in the generous creation of the world, something about the nature of God reveals itself. He is someone who wanted there to be a world that might freely come to know and love him, and he wants us to reproduce that love in our relations with one another.

In the basilica of San Clemente in Rome, there is a mosaic that reflects the concrete understanding of these truths. Above and behind the central altar, a crucified Christ sends out vines from the foot of the cross, which encircle John the Baptist and the doctors of the Church. The rivers of paradise flow from the same source to quench the thirst of stags and to water pastures and other parts of the creation. The pictorial message seems to be that the whole world emanates from, and is cradled in, that suffering love. The basilica itself stands on top of the ruins of imperial buildings and earlier churches and, at a level now several stories below the street, an ancient temple of Mithras. Unintentionally, but with a remarkable significance all the same, that medieval work of art reminds us that we are only one moment in an ongoing history that gathers into a universal story some of the insights of nonbiblical civilizations and religions along with the biblical tradition.

That story has no final ending in this life. In our interaction with the world, we are in very much the same situation as in our interaction with one another. We struggle with internal and external obstacles on our uncertain path towards the good, the true, and the beautiful. Our certain obligations towards our neighbors — now grown numerous by the very powers God and the world have built into us — must include obligations towards a nature without which we cannot flourish. We discover great challenges and uncertainties on every side. But that is part of the inspiring drama of human existence. And the outcome of that struggle is assured, even if we are not sure about it. The Creator who brought the world into being is certainly still at work within it and us.

Notes

INTRODUCTION

1. Debra West, "Good for Karma. Bad for Fish? Free Animals, Buddhists Say. Wildlife Experts Dissent," *New York Times,* January 11, 1997, 27.

2. Ibid., 31.

3. Stephen Bede Scharper, *Redeeming the Time: A Political Theology of the Environment* (New York: Continuum, 1997), 11-12.

4. Frederick Turner, "Cultivating the American Garden," in Cheryll Glotfelty and Harold From, eds., *The Ecocriticism Reader* (Athens, Ga.: University of Georgia Press, 1996), 40.

5. Al Gore, *Earth in the Balance* (New York: Houghton Mifflin, 1992), 19ff.

6. William K. Hartmann and Ron Miller, *The History of the Earth* (New York: Workman Publishing, 1991), 188-89.

7. T. H. Huxley, "The Origin of Species" (1860), in *Darwiniana Essays* (New York, 1893), 52.

8. See "Prayer to the Virgin of Chartres," in *Henry Adams: Novels, Mont Saint Michel, The Education* (New York: Library of America, 1983), 1202-7.

9. Henry Adams, *The Education of Henry Adams* (Boston: Houghton Mifflin, 1961), 384-85.

10. Roderick Nash, *Wilderness and the American Mind,* 3rd ed. (New Haven: Yale University Press, 1982), 138. On Muir in general, see Nash's whole chapter "John Muir: Publicizer," 122-40.

11. Ibid., 130.

12. On this point see Donald Worster, *The Wealth of Nature: Environmental History and the Ecological Imagination* (New York: Oxford University Press, 1993), esp. chap. 15, "John Muir and the Roots of American Environmentalism," 184-202; see also Robert H. Nelson, "Calvinism Minus God: Environmental Restoration as a Theological Concept," in L. Anthea Brooks and Stacy D. Van Deveer, eds., *Saving the Seas: Values, Scientists, and International Governance* (College Park: Maryland Sea Grant Book, 1997).

13. Nash, *Wilderness*, 128.

14. Jean Jacques Rousseau, *Emile*, translated by Allan Bloom (New York: Basic Books, 1979), 204.

15. Nash, *Wilderness*, 126.

16. For a good overview of this little-known school, see Terry L. Anderson and Donald R. Leal, *Free Market Environmentalism* (Boulder, Colo.: Westview Press, 1991).

17. Nash, *Wilderness*, 161.

18. Hans Urs von Balthasar, *A Theology of History* (New York: Sheed & Ward, 1963), 124-25.

19. The classic analysis of this process is John Henry Newman, *An Essay on the Development of Christian Doctrine* (South Bend: University of Notre Dame Press, 1989). The original essay appeared in 1888.

20. For a more nature-friendly analysis of Plato, see Gabriela Roxana Carone, "Plato and the Environment," *Environmental Ethics* 20 (Summer 1998).

21. See Edward Grant, "When Did Modern Science Begin?" *American Scholar* (Winter 1997): 105-13.

22. Among other works, see Stanley Jaki, *The Road of Science and the Ways to God* (Chicago: University of Chicago Press, 1987) and *God and the Cosmologists* (Washington, D.C.: Gateway Editions, 1989).

23. Francis Bacon, *Advancement of Learning*, Book I.

24. Kallistos Ware, "Ways of Prayer and Contemplation," in *Christian Spirituality*, edited by Brian McGinn and John Meyendorff (New York: Crossroads, 1985), 398.

25. Åke Hultkrantz, *Belief and Worship in Native North America*, edited by Christopher Vecsey (Syracuse, N.Y.: Syracuse University Press, 1981), 24.

26. William Denevan, "The Pristine Myth: The Landscape of the Americas in 1492," *Annals of the Association of American Geographers* 82 (3): 369-385.

27. Robert Royal, *1492 and All That: Political Manipulations of History* (Washington, D.C.: Ethics and Public Policy Center, 1992).

28. Sallie McFague, *The Body of God: An Ecological Theology* (Philadelphia: Fortress Press, 1993), 6.

29. J. C. Polkinghorne, "So Finely Tuned a Universe: Of Atoms, Stars, Quanta and God," *Commonweal*, August 16, 1996, 18.

30. Walker Percy, *Lost in the Cosmos: The Last Self-Help Book* (New York: Farrar, Straus & Giroux, 1983), 99.

CHAPTER ONE
"The Bible Made Me Do It?"

1. *Popol Vuh*, translated by Dennis Tedlock (New York: Touchstone/Simon & Schuster, 1996), 68.

2. Edward O. Wilson, *In Search of Nature* (Washington, D.C.: Shearwater Press, 1996), 184.

3. White's essay first appeared in *Science*, March 10, 1967, and has been reprinted in many places.

4. Ernest L. Fortin, "The Bible Made Me Do It: Christianity, Science, and the En-

vironment," in *Human Rights, Virtue, and the Common Good: Untimely Meditations on Religion and Politics* (New York: Rowman & Littlefield, 1997), 111-33.

5. The Holy Bible, Revised Standard Version.

6. The Holy Bible, New American Bible.

7. Gregg Easterbrook, *A Moment on the Earth* (New York: Penguin Books, 1995), xvii.

8. On the rights of trees, see Christopher Stone, *Should Trees Have Standing?* (Los Altos, Calif.: William Kaufmann, 1974). Stone later retreated from this position after seeing the excesses to which such notions were put. On rivers and rocks, see Roderick Frazier Nash, *The Rights of Nature* (Madison: University of Wisconsin Press, 1989).

9. Wilson, *In Search of Nature*, 176.

10. René Descartes, *Discourse on the Method of Rightly Conducting the Reason*, in *The Philosophical Works of Descartes*, vol. 1, translated by Elizabeth S. Haldane and G. R. T. Ross (New York: Cambridge University Press, 1973), 119.

11. The best introduction to this subject is Paul Santmire's *The Travail of Nature: The Ambiguous Ecological Promise of Christian Theology* (Philadelphia: Fortress Press, 1973).

12. All quotations from the *Confessions* are from the translation by J. G. Pilkington in Whitney J. Oates, ed., *Basic Writings of Saint Augustine*, vol. 1 (New York: Random House, 1948).

13. Abraham Joshua Heschel, *The Sabbath: Its Meaning for Modern Man* (New York: Farrar, Straus & Giroux, 1951), 8.

14. On Augustine and earlier notions of progress, see Robert Nisbet, *History of the Idea of Progress* (New York: Basic Books, 1980).

15. St. Augustine, *Basic Writings*, vol. 2, 202.

16. St. Augustine, *Epistle*, 138.1

17. Quoted in Michael Novak, *The Fire of Invention: Civil Society and the Future of the Corporation* (Lanham, Md.: Rowman & Littlefield, 1997), 56.

18. Santmire, *Travail of Nature*, 66.

19. St. Augustine, *Basic Writings*, vol. 2, 476-77.

20. Santmire, *Travail of Nature*, 70.

21. Cardinal Joseph Ratzinger, *"In the Beginning . . .": A Catholic Understanding of the Story of Creation and the Fall*, translated by Boniface Ramsey, O.P. (Grand Rapids, Mich.: Eerdmans, 1995), ix.

22. See, e.g., Will Durant, *The Story of Philosophy* (New York: Washington Square Press, 1970), 163.

23. Ratzinger, *"In the Beginning . . . ,"* 8.

24. Ibid., 12.

25. On this point and its importance to Christian evangelization in the New World, see Robert Royal, *1492 and All That: Political Manipulations of History* (Washington, D.C.: Ethics and Public Policy Center, 1992), 131ff.

26. Ratzinger, *"In the Beginning . . . ,"* 18.

27. Ibid., 23.

28. Ibid., 32.

29. Ibid., 35. Ratzinger may here be confusing Galileo with Bacon, though the common mentality of the time would probably have made the notion acceptable to either.

30. Ibid., 36.

31. Ibid., 45.

32. Christopher Derrick, *The Delicate Creation* (Old Greenwich: Conn.: Devin-Adair, 1972), 21.

33. Gerald Von Rad, *Genesis: A Commentary,* translated by J. H. Marks (Philadelphia: Westminster Press, 1972), 89.

34. Ratzinger, *"In the Beginning . . . ,"* 68.

35. Ibid., 69, 74.

36. Ibid., 82.

37. Ibid., 93.

38. Ibid., 96-97.

39. Ibid., 99-100.

40. Leszek Kolakowski, *Modernity on Endless Trial* (Chicago: University of Chicago Press, 1990), 7-8.

CHAPTER TWO
"A Dull Child's Guide to the Cosmos"

1. See the July 20, 1998, issue of these magazines.

2. For example, Daniel C. Dennett has argued that religious believers are like wild animals needing restraint and parents must be prevented from teaching their children false ideas about evolution. *Darwin's Dangerous Idea: Evolution and the Meanings of Life* (New York: Simon & Schuster, 1955), 515-16.

3. Edward J. Larson and Larry Witham have found in a sample of 1,000 randomly selected scientists that 39.3 per cent describe themselves as believers. In 1916, when a similar survey was conducted by James Leuba, 41.8 per cent responded positively. Larson and Witham also found that almost 15 per cent of scientists are agnostic on the question, so the nonbelievers (45.3 per cent) are only slightly more numerous than the believers. See Larson and Witham, "Scientists Are Still Keeping the Faith," *Nature,* April 3, 1997, 435-36.

4. Albert Einstein, *Out of My Later Years* (New York: Philosophical Library, 1950).

5. Thomas Aquinas, *Summa Contra Gentiles,* translated by Vernon J. Bourke (Notre Dame: University of Notre Dame Press, 1975), Book III, 1.

6. Ibid., Book III, chap. 69.

7. Aquinas, *Summa Theologiae,* II-II, ix, 4.

8. One of the best accounts of the universe's origins, though now beginning to age a bit, is Steven Weinberg, *The First Three Minutes: A Modern View of the Origin of the Universe* (New York: Basic Books, 1977).

9. Degrees Kelvin (K) correspond in size with the degrees in the scientific Celsius scale, but instead of setting zero at the freezing point of water, Kelvin's scale begins with "absolute zero," approximately -273° C, a point at which, theoretically, no energy remains in a physical system.

10. John Boslough, *Stephen Hawking's Universe* (New York: Avon Books, 1989), 109.

11. See the account with notes and the entire discussion of the Big Bang/Creator question by Ernan McMullin, "How Should Cosmology Relate to Theology?" in

The Sciences and Theology in the Twentieth Century, edited by A. R. Peacocke (Notre Dame: University of Notre Dame Press, 1981), 32ff.

12. William K. Hartmann and Ron Miller, *The History of Earth* (New York: Workman Publishing, 1991), 43.

13. On the orbital question, see Curt Covey, "The Earth's Orbit and the Ice Ages," *Scientific American,* February 1984, 58; on sunspot correlation with warming, see Daniel B. Botkin and Edward A. Keller, *Ecological Science: Earth as a Living Planet* (New York: Wiley, 1995), 441.

14. Paul Davies, *God and the New Physics* (New York: Touchstone, 1983), 179.

15. Roger Penrose, *The Emperor's New Mind: Concerning Computers, Minds, and the Laws of Physics* (Oxford: Oxford University Press, 1989), 344. Cited in Richard Denton, *Biology: The Anthropic Perspective* (unpublished typescript), chap. 2, p. 2.

16. Boslough, *Stephen Hawking's Universe,* 109.

17. Quoted in Michael White and John Gribbin, *Stephen Hawking: A Life in Science* (New York: Dutton, 1992), 201.

18. Stephen W. Hawking, *A Brief History of Time: From the Big Bang to Black Holes* (Toronto: Bantam, 1988), 174.

19. For popular accounts of quantum theory, see Davies, *God and the New Physics;* or at a slightly more advanced level, J. C. Polkinghorne, *The Quantum World* (Princeton: Princeton University Press, 1989). The bibliographies in both volumes are a sound guide to the primary sources by the great theorists in the development of quantum theory such as Max Planck, Niels Bohr, Paul Dirac, and Werner Heisenberg.

20. Polkinghorne, *Quantum World,* 97.

21. See J. C. Polkinghorne's *Quarks, Chaos, and Christianity: Questions to Science and Religion* (New York: Crossroad, 1996).

22. Boslough, *Stephen Hawking's Universe,* 115.

23. Dennett, *Darwin's Dangerous Idea.*

24. Dean L. Overman, *A Case Against Accident and Self-Organization* (Lanham, Md.: Rowman & Littlefield, 1997).

25. Michael J. Behe, *Darwin's Black Box: The Biochemical Challenge to Evolution* (New York: Free Press, 1996), 69-73.

26. See the famous article by F. H. C. Crick and L. E. Orgel, "Directed Panspermia," *Icarus* 19 (1973): 344.

27. Hartmann and Miller, *History of Earth,* 214-15.

28. Gregg Easterbrook, *A Moment on the Earth* (New York: Penguin Books, 1995), 25.

29. James E. Lovelock, *Gaia: A New Look at Life on Earth* (New York: Oxford University Press, 1979).

30. Easterbrook, *Moment on the Earth,* 690-91.

31. For a persuasive theory, see Doug McInnis, "And the Waters Prevailed," *Earth,* August 1998, 48-54.

32. On this and other quite interesting ways that geology may help explain otherwise inexplicable human records of natural events, see Dorothy B. Vitaliano, *Legends of the Earth: Their Geologic Origins* (Bloomington: Indiana University Press, 1973). For the Flood: 153ff.

33. Ibid., 144.

34. Ibid., 179-201.

35. Daniel B. Botkin, *Discordant Harmonies: A New Ecology for the Twenty-First Century* (New York: Oxford University Press, 1990), 6.

36. Ibid., 9, 8.

37. Ibid., 47.

38. Ibid., 62.

39. Quoted ibid., 63.

40. Andrew Brennan, *Thinking about Nature* (Athens: University of Georgia Press, 1988), 104.

41. See Daniel B. Botkin and Edward A. Keller, *Ecological Science: Earth as a Living Planet* (New York: Wiley, 1995), 160.

42. William Denevan, "The Pristine Myth: The Landscape of the Americas in 1492," *Annals of the Association of American Geographers* 82 (3): 369-85.

43. Botkin, *Discordant Harmonies*, 68.

44. Ibid., 89.

45. Ibid., 189.

46. Easterbrook, *Moment on the Earth*, 80-82.

47. William H. Calvin, "The Great Climate Flip-Flop," *Atlantic Monthly*, January 1998, 47.

48. M. E. Raymo et al., "Millennial-scale Climate Instability During the Early Pleistocene Epoch," *Nature*, April 16, 1998, 699-702.

49. Calvin, "Great Climate Flip-Flop," 64.

A HOPEFUL INTERLUDE

1. Fred Smith, "Reappraising Humanity's Challenges, Humanity's Opportunities," in Ronald Bailey, ed., *The True State of the Planet* (New York: Free Press, 1995), 379.

2. Al Gore, "An Ecological Kristallnacht. Listen," *New York Times*, March 19, 1989.

3. Unfortunately, many programs in schools do not take a tempered, informed approach to training environmentally sensitive students. For an analysis of what is good and bad in environmental teaching, see Michael Sanera and Jane S. Shaw, *Facts Not Fear: A Parent's Guide to Teaching Children about the Environment* (Washington, D.C.: Regnery, 1996).

4. See Michael Specter, "Population Implosion Worries a Graying Europe," *New York Times*, July 10, 1998, A1.

5. See United Nations, *World Population Prospects: The 1994 Revision* (New York: U.N. Department for Economic and Social Information and Policy Analysis, 1994).

6. Paul E. Waggoner, "How Much Land Can Be Spared for Nature?" *Daedalus* 125, no. 3 (Summer 1996): 87.

7. At the time of this writing, the most recent version is *State of the World 1994* (Washington, D.C.: Worldwatch Institute, 1994).

8. Lester R. Brown, Christopher Flavin, and Sandra Postel, *Saving the Planet: How to Shape an Environmentally Sustainable Global Economy* (New York: Norton, 1991). See chap. 6, "Grain for Eight Billion."

9. Stephen R. Graybeard, "Preface," *Daedalus* 125, no. 3 (Summer 1996): vi.

10. Ibid., vii-viii.

11. Jesse H. Ausubel, "The Liberation of the Environment," *Daedalus* 125, no. 3 (Summer 1996): 1.

12. For a careful review of the literature that comes to a similar conclusion, see Gregg Easterbrook, *A Moment on the Earth: The Coming Age of Environmental Optimism* (New York: Penguin, 1996), 246-48

13. Ausubel, "Liberation of the Environment," 14-15.

14. Easterbrook, *Moment on the Earth,* 162.

15. Norman Myers and Julian Simon, *Scarcity or Abundance?: A Debate on the Environment* (New York: Norton, 1994). See also E. O. Wilson, ed., *Biodiversity* (Washington, D.C.: National Academy Press, 1988), and Aaron Wildavsky, *Searching for Safety* (New Brunswick, N.J.: Transaction Press, 1988).

16. Easterbrook, *Moment on the Earth,* 559.

17. The figures represent estimates by the World Resources Institute in *World Resources 1994-5* (New York: Oxford University Press, 1994).

18. *Science,* January 24, 1997.

19. William K. Stevens, " 'Hot Spots' for American Endangered Species Cover Surprisingly Little Land," *New York Times,* January 24, 1997, A22.

20. Al Gore, *Earth in the Balance* (New York: Houghton Mifflin, 1992), 85.

21. In 1998, several scientists who believe anthropogenic global warming is occurring pointed out flaws in the satellite data owing to orbital decay. Roy W. Spencer, one of the designers of the satellite system at NASA's Marshall Space Flight Center, has conceded that the cooling is less than previously thought—0.01 degrees Celsius per decade—but is still there. Roy Spencer, "When Science Meets Politics," *Washington Times,* September 3, 1998, A19.

CHAPTER THREE
"Back and Forth"

1. Al Gore, *Earth in the Balance* (New York: Houghton Mifflin, 1992), 79.

2. Thomas Berry, *The Dream of the Earth* (San Francisco: Sierra Club Books, 1990), 18.

3. Ibid., 10.

4. Brian Swimme and Thomas Berry, *The Universe Story: From the Primordial Flaring Forth to the Ecozoic Era, A Celebration of the Unfolding of the Cosmos* (San Francisco: Harper Collins, 1992).

5. Berry, *Dream of the Earth,* xiv.

6. Thomas Berry, "Economics: Its Effect on the Life Systems of the World," in Anne Lonergan and Caroline Richards, eds., *Thomas Berry and the New Cosmology* (Mystic, Conn.: Twenty-third Publications, 1987), 25.

7. Berry, *Dream of the Earth,* 35.

8. Ibid., 33.

9. Ibid., 209.

10. Swimme and Berry, *Universe Story,* 243.

11. For instance, his essay "The Viable Human" leads off the opening section

("What Is Deep Ecology?") of the highly important anthology *Deep Ecology for the 21st Century*, edited by George Sessions (Boston: Shambala, 1995), 8-18.

12. Swimme and Berry, *Universe Story*, 2.

13. Ibid., 74.

14. Ibid., 76-77.

15. Ibid., 77.

16. Ibid., 78.

17. Ibid., 59.

18. Ibid., 15.

19. Quoted in Robert Whelan, Joseph Kirwan, and Paul Haffner, *The Cross and the Rain Forest: A Critique of Radical Green Spirituality* (Grand Rapids, Mich.: Acton/Eerdmans, 1996), 31. The original statement appears in Thomas Berry and Thomas Clarke, *Befriending the Earth: A Theology of Reconciliation between Humans and the Earth* (Mystic, Conn.: Twenty-third Publications, 1991), 76.

20. Swimme and Berry, *Universe Story*, 140.

21. Ibid., 175.

22. Ibid., 12-14.

23. Ibid., 56.

24. Ibid., 14.

25. See, e.g. Berry, *Dream of the Earth*, 79; and Lonergan and Richards, *Thomas Berry and the New Cosmology*, 14.

26. See Thomas Aquinas, *Summa Theologiae*, I-II q. 105 aa. 2-3, and II-II q. 66.

27. Berry, *Dream of the Earth*, 37.

28. Ibid., 77.

29. Virginia I. Postrel, "Laissez Fear," *Reason Magazine*, April 1997, 4.

30. See Stephen Bede Scharper, *Redeeming the Time: A Political Theology of the Environment* (New York: Continuum, 1997), 129.

31. Swimme and Berry, *Universe Story*, 250.

32. Ibid., 58.

33. Ibid.

34. Berry, *Dream of the Earth*, 60-61.

35. Frederick Turner, *The Culture of Hope: A New Birth of the Classical Spirit* (New York: Free Press, 1995).

36. Frederick Turner, "Escape from Modernism: Technology and the Future of the Imagination," *Harper's*, November 1984, 49.

37. Turner, *Culture of Hope*, 82.

38. Ibid., 7.

39. Ibid., 5.

40. Ibid., 5-6.

41. Ibid., 77.

42. Ibid., 79.

43. Ibid., 80.

44. Ibid., 89.

45. Ibid., 85-90.

46. Ibid., 11.

47. Ibid., 72.

48. Ibid., 83.

49. Ibid.
50. Ibid., 21.
51. Ibid., 120.
52. Ibid., 102.
53. Ibid., 100.
54. Ibid., 81.
55. Ibid., 241.
56. Ibid., 28.
57. Ibid., 83.
58. Ibid., 218.
59. Ibid., 223.
60. Ibid., 231.

61. Romano Guardini, *Letters from Lake Como: Explorations in Technology and the Human Race* (Grand Rapids, Mich.: Eerdmans, 1994), 96. These letters tell the whole story of Guardini's developing notions about nature and technology.

Chapter Four
"Deep, Deeper, Deepest"

1. George Sessions, ed., *Deep Ecology for the Twenty-first Century* (Boston: Shambala, 1995), ix-x.

2. Arne Naess, "Deep Ecology in Good Conceptual Health," *Trumpeter* 3, no. 4 (Fall 1986): 20.

3. Naess has typically expressed his views in short essays or sketches. The most substantial presentation of his position is Arne Naess, *Ecology, Community, and Life-style: Outline of an Ecosophy,* translated by David Rothenberg (New York: Cambridge University Press, 1989). For an overview of the Deep Ecology movement, see Sessions, *Deep Ecology.*

4. See, for just one instance, Arne Naess, "Deepness of Questions," in Sessions, *Deep Ecology,* 210.

5. These points have been reproduced in many places. A convenient place to find them, with a discussion of their significance, is Arne Naess, "The Deep Ecological Movement: Some Philosophical Aspects," in Sessions, *Deep Ecology,* 64-84 (platform on 68).

6. George Sessions, "Ecological Consciousness and Paradigm Change," in Michael Tobias, ed., *Deep Ecology* (San Diego: Avant Books, 1985), 30.

7. See, e.g., David Ehrenfeld, *The Arrogance of Humanism* (New York: Oxford University Press, 1978), 259-60.

8. For the two quotations, see Al Gore, *Earth in the Balance: Ecology and the Human Spirit* (New York: Houghton Mifflin, 1992), 217.

9. Lynn White, Jr., "The Future of Compassion," *Ecumenical Review* 30, no. 2 (April 1978): 108.

10. Sessions, *Deep Ecology,* xiii.

11. Arne Naess, "The Deep Ecology 'Eight Points' Revisited," in Sessions, *Deep Ecology,* 213, 217.

12. In Sessions, *Deep Ecology,* 28.

13. Bill Devall, *Simple in Means, Rich in Ends: Practicing Deep Ecology* (Salt Lake City: Gibbs Smith, 1988), 49.

14. Arne Naess, "Self-Realization: An Ecological Approach to Being in the World," in Sessions, *Deep Ecology,* 239.

15. Michael L. Zimmerman, *Contesting Earth's Future: Radical Ecology and Postmodernity* (Berkeley and Los Angeles: University of California Press, 1994), 59.

16. Richard Langlais, "Living in the World: Mountain Humility, Great Humility," in Sessions, *Deep Ecology,* 196.

17. Arne Naess, "The Place of Joy in a World of Fact," in Sessions, *Deep Ecology,* 250.

18. Naess, "Deep Ecological Movement," 80.

19. For a history of Spinoza's problems with the Amsterdam Jewish community and Bergson's remark, see Yirmiahu Yovel, "Why Spinoza Was Excommunicated," *Commentary,* November 1977, 46-52.

20. Arne Naess, *Freedom, Emotion, and Self-Subsistence: The Structure of a Central Part of Spinoza's Ethics* (Oslo: Universitetsforlaget, 1975), 118-19.

21. On these two points, see Zimmerman, *Contesting Earth's Future,* 41.

22. See Henryk Skolimowski, "The Dogma of Anti-Anthropocentrism and Ecophilosophy," *Environmental Ethics* 6 (Fall 1984): 287.

23. Charles T. Rubin, *The Green Crusade: Rethinking the Roots of Environmentalism* (New York: Free Press, 1994), 184.

24. Recounted in Zimmerman, *Contesting Earth's Future,* 45.

25. Curtis L. Hancock, "Philosophers in the Mist," *Crisis,* March 1996, 38-43.

26. Paul Johnson, *The Quest for God: A Personal Pilgrimage* (New York: HarperCollins, 1996), 84.

27. Ibid., 91.

28. Bill Devall and George Sessions, *Deep Ecology: Living as if Nature Mattered* (Salt Lake City: Peregrine Smith, 1985), 201, quoted in Rubin, *Green Crusade,* 291.

29. Devall, *Simple in Means,* 39, quoted in Rubin, *Green Crusade,* 204.

30. Martin W. Lewis, *Green Delusions: An Environmentalist Critique of Radical Environmentalism* (Durham: Duke University Press, 1992), 29.

31. Naess, *Ecology, Community, and Lifestyle,* 211.

32. *Catechism of the Catholic Church* (Washington, D.C.: U.S. Catholic Conference, 1994), sec. 2418.

33. Ibid., secs. 2417-8.

CHAPTER FIVE
"The Gospel According to Matthew"

1. For instance, see the papal statement and another by the U.S. Catholic Conference in *"And God Saw That It Was Good": Catholic Theology and the Environment,* Drew Christiansen and Walter Grazer, eds. (Washington, D.C.: U.S. Catholic Conference, 1996).

2. Editorial, "Silencing Fox a Fruitless Exercise," *National Catholic Reporter,* October 21, 1988, 10.

3. For Chenu's analysis, see M. D. Chenu, *Nature, Man, and Society in the Twelfth Century* (Chicago: University of Chicago Press, 1968). For Fox's commentary on this

material, see his introduction to *Sheer Joy: Conversations with Thomas Aquinas on Creation Spirituality* (San Francisco: HarperCollins, 1992).

4. Matthew Fox and Rupert Sheldrake, *Natural Grace: Dialogues on Creation, Darkness, and the Soul in Spirituality and Science* (New York: Image, 1997).

5. Matthew Fox, *The Coming of the Cosmic Christ* (San Francisco: HarperSanFrancisco, 1988), 57.

6. For some other ways of thinking about panentheism, see Richard John Neuhaus, "Christ and Creation's Longing," in Thomas Sieger Derr, *Environmental Ethics and Christian Humanism* (Nashville: Abingdon Press, 1998), 124.

7. Fox, *Coming of the Cosmic Christ*, 145.

8. Matthew Fox, *Creation Spirituality: Liberating Gifts for the Peoples of the Earth* (San Francisco: HarperSanFrancisco, 1991), 16.

9. Ibid., 13.

10. Fox, *Coming of the Cosmic Christ*, 234.

11. Fox, *Creation Spirituality*, 25.

12. Matthew Fox, *The Reinvention of Work: A New Vision of Livelihood for Our Time* (New York: HarperCollins, 1997).

13. Ibid., 221.

14. These suggestions are sprinkled throughout *Original Blessing: A Primer in Creation Spirituality, presented in Four Paths, Twenty-Six Themes, and Two Questions* (Santa Fe, N.M.: Bear, 1983).

15. Fox, *Coming of the Cosmic Christ*, 15.

16. Quoted in Donna Steichen, *Ungodly Rage* (San Francisco: Ignatius Press, 1992), 226. Steichen's volume is the only attempt to trace the enormous influence Fox and others have had on different religious communities, church education programs, and general attitudes in the United States.

17. Ibid., 42.

18. R. C. Anderson, "The Historic Role of Fire in the North American Grassland," in S. L. Collins and L. L. Wallace, eds., *Fire in North American Tallgrass Prairies* (Norman: University of Oklahoma Press, 1990), 14.

19. Fox and Sheldrake, *Natural Grace*, 4.

20. In his introduction to *The City of God,* Thomas Merton, who knew something about mysticism and the spiritual life, begins with the observation: "Here is a book that was written over fifteen hundred years ago by a mystic in North Africa." *City of God,* translated by Marcus Dods (New York: The Modern Library, 1950), ix.

CHAPTER SIX
"Sophia's World"

1. A basic introduction to several dimensions of this movement may be found in Greta Gaard, ed., *Ecofeminism: Women, Animals, Nature* (Philadelphia: Temple University Press, 1993); and Irene Diamond and Gloria Feman Orenstein, eds., *Reweaving the World: The Emergence of Ecofeminism* (San Francisco: Sierra Club Books, 1990).

2. For a fairly typical example of how the Virgin is even conflated with non-Christian goddesses for radical purposes, see Sandra Cisneros, "Guadalupe the Sex

Goddess: Unearthing the Racy Past of Mexico's Most Famous Virgin," *Ms. Magazine,* July/August 1996, 43-46.

3. See Elaine Pagels, *The Gnostic Gospels* (New York: Random House, 1979).

4. Marija Gimbutas, *The Goddesses and Gods of Old Europe: 6500-3500 B.C.* (Berkeley and Los Angeles: University of California Press, 1982).

5. The expression is widely used, but see especially Karen Warren, "The Power and Promise of Ecological Feminism," *Environmental Ethics* 12 (1990): 125-47.

6. Bernice-Marie Daly, *Ecofeminism: Sacred Matter/Sacred Mother* (Chambersburg, Pa.: Anima Books, published for the American Teilhard Association for the Future of Man, 1991), 3. For a full historical view of this alleged identity between antifemale attitudes and technological damage to the earth, see Carolyn Merchant, *The Death of Nature: Women, Ecology, and the Scientific Revolution* (New York: Harper & Row, 1980).

7. Margaret Brennan, "Patriarchy: The Root of the Alienation from the Earth," in Anne Lonergan and Caroline Richards, eds., *Thomas Berry and the New Cosmology* (Mystic, Conn.: Twenty-third Publications, 1987), 57.

8. Claude Masset, "Prehistory of the Family," in André Burguière et al., eds., *A History of the Family,* vol. 1, *Distant Worlds, Ancient Worlds* (Cambridge: Harvard University Press, 1996), 76.

9. Françoise Zonabend, "An Anthropological Perspective on Kinship and the Family," in Burguière, *History of the Family,* vol. 1, 43-44.

10. Janet Biehl, *Rethinking Ecofeminist Politics* (Boston: South End Press, 1991), 40.

11. Ibid., 32.

12. Ibid., 49.

13. Ibid., 43.

14. Rene Denfeld, *The New Victorians: A Young Woman's Challenge to the Old Feminist Order* (New York: Warner, 1995), 128.

15. Ibid, 128-9.

16. Ibid., 143.

17. Ibid., 141.

18. Grace Jantzen, "Healing Our Brokenness: The Spirit and the Creation," *Ecumenical Review* 42, no. 2 (1990): 132.

19. Carol P. Christ, "Rethinking Theology and Nature," in Diamond and Orenstein, eds., *Reweaving the World,* 58-69.

20. Susan Griffin, *Woman and Nature: The Roaring Inside Her* (New York: Harper & Row, 1978), xvii.

21. Rosemary Radford Ruether, *Gaia and God: An Ecofeminist Theology of Earth Healing* (San Francisco: HarperCollins, 1992).

22. Rosemary Radford Ruether, "Goddesses and Witches: Liberation and Counter-cultural Feminism," *The Christian Century,* September 10-17, 1980, 842. Quoted in Donna Steichen, *Ungodly Rage: The Hidden Face of Catholic Feminism* (San Francisco: Ignatius Press, 1991), 35. Although Steichen's book contains some remarkable material about events at which ecofeminism has been promoted, and although she probably understands the origins and development of the whole movement better than any other analyst, polemical passion sometime leads her into overstatements. The whole book needs to be read carefully and critically.

23. Michael S. Rose, "Feminist Theologian Urges Religious to Find Way to 'Weed Out People,' " *The Wanderer,* June 11, 1998, 1.

24. Elizabeth Johnson, "The Cosmos: An Astonishing Image of God," reprinted in *Origins* 26 (September 12, 1996): 206-7.

25. Marti Kheel, "Ecofeminism and Deep Ecology: Reflections on Identity and Difference," in Diamond and Orenstein, *Reweaving the World*, 128-37.

26. Remarks on the value of hunting in Deep Ecology may be found in one of the key texts of that movement: Bill Devall and George Sessions, *Deep Ecology: Living as if Nature Mattered* (Salt Lake City: Peregrine Smith, 1985).

27. Cited in Devall and Sessions, *Deep Ecology*, 137. The original phrase is from Charlene Spretnak, ed., *The Politics of Women's Spirituality* (New York: Anchor Books, 1982), xvii.

28. Johnson, "The Cosmos," 210.

29. Ruether, *Gaia and God*, 268.

30. Simone de Beauvoir, *The Second Sex* (New York: Vintage, 1974).

31. See e.g., Vandana Shiva, "Let Us Survive: Women, Ecology, and Development," in Rosemary Radford Ruether, ed., *Women Healing Earth: Third World Women on Ecology, Feminism, and Religion* (Maryknoll, N.Y.: Orbis Books, 1996), 65-73. For a powerful critique of the idea of "gendered" science and other attempts to debunk science as socially constructed, see Paul R. Gross and Norman Levitt, *Higher Superstition: The Academic Left and Its Quarrels with Science* (Baltimore: Johns Hopkins University Press, 1994).

32. See Deal W. Hudson's enlightening treatment of this and other questions, "Human Nature, Gender, and Ethnicity," in *The Great Books Today* (Chicago: Encyclopedia Britannica, 1995), 127-67.

33. Ruether, *Women Healing Earth*, 5.

34. Ibid., 10-11.

35. Francis Martin, *The Feminist Question: Feminist Theology in the Light of Christian Tradition* (Grand Rapids, Mich.: Eerdmans, 1994).

CHAPTER SEVEN
"Liberation and Its Discontents"

1. Quoted in Thomas Sieger Derr, *Ecology and Human Liberation: A Theological Critique of the Use and Abuse of Our Birthright* (New York: WSCF Books, 1973), 95.

2. Quoted in Gregg Easterbrook, *A Moment on the Earth: The Coming Age of Environmental Optimism* (New York: Penguin, 1995), 600.

3. Ibid., 570, 580.

4. Ibid., 583.

5. Karl Marx, *The German Ideology*, in Robert C. Tucker, ed., *The Marx-Engels Reader* (New York: Norton, 1972), 133.

6. Leonardo Boff, *Cry of the Earth, Cry of the Poor*, translated by Phillip Berryman (Maryknoll, N.Y.: Orbis Books, 1997), xi.

7. Ibid., 13.

8. Ibid., 63.

9. Ibid., 65.

10. Ibid., 66.

11. Ibid., 66-67.

12. Ibid., 89.

13. Ibid., 90.

14. Ibid., 100-101.

15. Allen Johnson, "How the Machihuenga Manage Resources: Conservation or Exploitation of Nature?" in A. Posey and W. Baloe, eds., *Advances in Economic Botany* (New York: New York Botanical Garden, 1989), 221, quoted in Wallace Kaufman, *No Turning Back: Dismantling the Fantasies of Environmental Thinking* (New York: Basic Books, 1994), 61.

16. Easterbrook, *Moment on the Earth,* 420.

17. Boff, *Cry of the Earth,* 111.

18. Martin W. Lewis, *Green Delusions: An Environmental Critique of Radical Environmentalism* (Durham: Duke University Press, 1992), 15.

19. Ibid., 7.

20. Ibid., 9.

21. Ibid.

22. Ibid., 11.

23. Ibid., 135.

24. Ibid., 10, 19.

25. Quoted ibid., *Green Delusions,* 195.

26. Ibid., 197.

27. Ibid., 201.

28. Ibid., 2.

29. Al Gore, *Earth in the Balance: Ecology and the Human Spirit* (Boston: Houghton Mifflin, 1992), 259, 14.

30. Ibid., 274.

31. Ibid., 30.

32. Ibid., 260.

33. Ibid., 257.

34. Ibid., 163, 144, 244.

35. Ibid., 223-26.

36. Ibid., 241.

37. Ibid., 364.

38. Ibid., 207.

39. Derr, *Ecology and Human Liberation,* 33.

40. Ibid., 35.

41. Ibid., 36.

42. Ibid., 39.

43. Ibid., 44.

44. Thomas Sieger Derr, *Environmental Ethics and Christian Humanism,* with critical responses by James A. Nash and Richard John Neuhaus (Nashville: Abingdon Press, 1998), 61.

45. "Joint Appeal by Religion and Science for the Environment," in *BioScience* 41, no. 8 (September 1992): 625.

46. Derr, *Environmental Ethics,* 94.

Index of Names

NOTE: This index of proper names also includes a few important terms that are not capitalized in the text: big bang, biocentrism, biodiversity, bioregionalism, chlorofluorocarbons, rights of nature.